FINE ARTS, MUSIC AND LITERATURE

APPLICATIONS IN ARTS BIOMECHANICS

FINE ARTS, MUSIC AND LITERATURE

Additional books in this series can be found on Nova's website
under the Series tab.

Additional e-books in this series can be found on Nova's website
under the e-book tab.

FINE ARTS, MUSIC AND LITERATURE

APPLICATIONS IN ARTS BIOMECHANICS

GONGBING SHAN
AND
PETER VISENTIN
EDITORS

nova publishers
New York

For permission to use material from this book please contact us:
Telephone 631-231-7269; Fax 631-231-8175
Web Site: http://www.novapublishers.com

NOTICE TO THE READER

The Publisher has taken reasonable care in the preparation of this book, but makes no expressed or implied warranty of any kind and assumes no responsibility for any errors or omissions. No liability is assumed for incidental or consequential damages in connection with or arising out of information contained in this book. The Publisher shall not be liable for any special, **consequential, or exemplary damages resulting, in whole or in part, from the readers' use of,** or reliance upon, this material. Any parts of this book based on government reports are so indicated and copyright is claimed for those parts to the extent applicable to compilations of such works.

Independent verification should be sought for any data, advice or recommendations contained in this book. In addition, no responsibility is assumed by the publisher for any injury and/or damage to persons or property arising from any methods, products, instructions, ideas or otherwise contained in this publication.

This publication is designed to provide accurate and authoritative information with regard to the subject matter covered herein. It is sold with the clear understanding that the Publisher is not engaged in rendering legal or any other professional services. If legal or any other expert assistance is required, the services of a competent person should be sought. FROM A DECLARATION OF PARTICIPANTS JOINTLY ADOPTED BY A COMMITTEE OF THE AMERICAN BAR ASSOCIATION AND A COMMITTEE OF PUBLISHERS.

Additional color graphics may be available in the e-book version of this book.

Library of Congress Cataloging-in-Publication Data

ISBN: 978-1-62808-015-5

Published by Nova Science Publishers, Inc. † New York

CONTENTS

PREFACE

Multidisciplinary has become catch-word of some currency. Notwithstanding the term's apparently increasing popularity, individuals who engage in multidisciplinary research typically find it to be considerably more difficult than they had initially anticipated. This is particularly true for multidisciplinary work involving disciplines founded on disparate fundamental research precepts. Some reasons for this are understandable. Often, from unidisciplinary perspectives, multidisciplinary research can seem to approach discourses laterally; as a result, both its undertaking and dissemination become challenging. Consequently, researchers in such collaborations frequently must concede the reformulation of their ideas to conform to one or another unidisciplinary expectation. In so doing, unfortunately, they can lose some of the significance of their message. Yet another solution to this challenge may be to engage in collaborations that coordinate only closely related disciplines. Although multidisciplinary in name, such collaborations are inherently designed to default to boundaries delineated by disciplinary canon, limiting both the questions that may be asked and how they might be examined.

In the research collectively represented in *Applications in Arts Biomechanics,* researchers sometimes from highly disparate disciplines use scientific technologies and methodologies to ask questions about the visual and performance arts. The work underscores some of the fundamental challenges of multidisciplinary research including the following: as research methodologies of constituent disciplines diverge, demands placed on researchers require that their efforts become complementary and integrative rather than merely coordinative. Perhaps this points toward the next stage in the evolution of multidisciplinary research, one where researchers supersede disciplinary boundaries, resulting in modes of thinking that become transdisciplinary.

For those researchers who try to negotiate a path that includes both science and art as part of an underlying research motivation, it remains a fundamental requirement to engage in clearly defined and rigorous research. Such work can challenge disciplinary dogma, whether that of the sciences or of the arts, through research that explores both Boolean intersections and liminal spaces defined by the boundaries of disciplinary thinking. This book is dedicated to explorations of intersections among the arts, sciences, and the humanistic nature of creative & scientific thinking. Its chapters investigate traditional biomechanics and the performance arts, performance pedagogy related to motor skills, human-object interaction, as well as biological and mechanical interfaces affecting the creative process. Such work may have broader vocational implications in the areas of ergonomics, locomotion, motor learning, technique optimization, design/redesign of performance tools, and vocational health & wellness research related to human performance. Simply stated, *Applications in Arts Biomechanics* aims to promote research that fosters links between creative and scientific modes of understanding.

Applications in Arts Biomechanics contains ten chapters covering music, dance, and the visual & media arts. It is organized in four parts: original research in each of the above plus a review article section. Music research covers a range of topics, from investigations correlating musical and motor development in children to a newly proposed methodology for the purpose of revealing relationships between gross and fine motor control during music performance. Dance articles examine the biomechanics of dancing and identify new ways of determining kinematic and kinetic factors that minimize technological interventions. Art and media chapters discuss in general the influence of technology **on "creation" in the visual and media** arts and also document a concrete example of one such application. Collectively, the Editors hope, these articles will encourage discussion and debate about the future of multidisciplinary research that explores the seemingly divergent disciplinary cultures of the arts and the sciences. Most importantly we hope that it will inspire new collaborations in Arts Biomechanics research, ones that aim toward transdisciplinary significance.

I – Music

In: Applications in Arts Biomechanics
Editors: Gongbing Shan and Peter Visentin
ISBN: 978-1-62808-015-5

Chapter 1

FOR 'LIVELY' AND 'INTENSELY RELATED' TONES IN BOWED STRING INSTRUMENTAL PERFORMANCE: A RESPONSE TO PETER RÖBKE'S INTRODUCTION TO *DAS MUSIZIEREN UND DIE GEFÜHLE*

Julia von Hasselbach
University of Freiburg, Germany

ABSTRACT

This article is a slightly changed English version of "Für, lebendige' und, beziehungsfähige' Töne im Streichinstrumentspiel. Eine Reaktion auf Peter Röbkes Einleitung zu *Das Musizieren und die Gefühle*," published in the proceedings of the symposium "*Musik und Gefühl. Interdisziplinäre Annäherungen in musikpädagogischer Perspektive*," Martina Krause and Lars Oberhaus (Eds.), Hildesheim (Olms)/ Germany 2012, pp. 67-88.

In response to a publication of the Viennese music educator Peter Röbke, who assumes a lack of consideration for the emotional lives of music students within the framework of their education and, consequentially, the musicians' inability to interconnect music and emotions during musical performances, this speculation is taken up and answered from the viewpoint of biomechanical research in the field of bowed string instrumental performance. A compilation of biomechanical criteria for good, but monotonous, sound production is complemented by criteria for differentiating sound formation, which may be more capable of transporting 'emotional messages' to the audience. A retrospective glance at important research results in 'strings biomechanics' is provided beforehand, to facilitate understanding of the article's argument for an instrumental education that gives empirical research results central consideration while constructing the noted concept of bowing 'technique.'

1. Background and Aims

The well-known music educator Peter Röbke[1] and psychoanalyst Helmuth Figdor[2], both currently teaching in Vienna,
have authored a pioneering book in the research[3] field of music and emotions. Röbke opens his introduction to *Music-Making and Emotions: Instrumental Education and Psychoanalysis in Dialogue*[4] with an essayistic approach: Listening to a violinist in concert at a music university, he considers the performance from a very personal, yet representative viewpoint. Although he labels the performance "technically perfect" (Röbke 2008, 10), Röbke states that he senses a number of deficits; for example, the absence of dynamic variety, of increased and decreased tension, of articulation and variation in timbre and vibrato. In conclusion, Röbke describes his impression that this violinist presented an "uninterrupted, uniform stream of tones" (ibid.) while "neglecting to give soul to the event of sound" (ibid.). Disappointed, Röbke speculates about a possibly causal "déformation professionelle" [French in the original] (ibid. 12) within the framework of education, before getting to the heart of the subject matter in various ways: What if the "emotional lives" (ibid.) of music students found little consideration within the development of their skills, so that sometimes music considered to be "communication of deep emotions through tones" would be reduced to "tones, bright and shiny as the stars, which have already been extinguished inside" (ibid.)?

This speculation is taken up in the following article, and enhanced by perspectives within an empirical research field positioned between instrumental education, sound physics, and biomechanics or music physiology, respectively. This 'inter-discipline' recently found international attention, thanks to the journal *Arts Biomechanics*[5], which publishes interdisciplinary research on biomechanical aspects of all fine arts.

This article hypothesises that the monotony in the 'perfection' of the described violinist is at least partially caused by a lack of consideration for, or a lack of education in, essential knowledge about string instrumental performance – with regard to the interrelation between the aspects of artistic objectives, sound physical conditions of the instrument, and biomechanical possibilities, as substantiated by empirical research results. The 'liveliness' and 'intensity of relationships' of the tones that Röbke found absent could, therefore, be cultivated – beyond the approaches presented by Röbke and Figdor – by a scientifically substantiated concept of violin 'technique'.[6]

With Röbke's introduction in mind – and as a reaction to it – the following contribution will demonstrate, through examples, what must have happened at the level of sound physics and biomechanics during this violinist's performance, to make it so monotonous to Röbke; and what should happen to increase the probability of the performance being received as

[1] Peter Röbke is director of the Institute for Music Education at the University for Music and Performing Arts Vienna: http://www.musiceducation.at/das-institut/personen/roebke-peter.

[2] Helmuth Figdor holds, among other qualifications, a postdoctoral degree in psychoanalytic education: http://www.app-wien.at/Figdor.

[3] Research in the humanities

[4] To the best of the author's knowledge, all citations have been accurately translated. Please refer to the original in cases of doubt. Cf. Röbke 2008, 9ff.

[5] See Shan and Visentin 2011

[6] This concept is to be understood particularly in contrast to the concept that 'technique' sorts itself out at some point, for example while intensely studying scales and etudes, without conscious cognitive learning of sound physical and biomechanical processes and their targeted optimization.

distinctly refined with respect to 'emotions', or as conveying 'emotional messages'.[7] For this purpose, criteria of particular relevance to the respective artistic results are compiled. Furthermore, some dangers of systematic failure of artistic expressive efforts are revealed.

Before trying to illustrate the relevance of sound physical and biomechanical research results in the instance of string instrumental performance, as shown at the symposium "Music and Emotion"[8], this article will facilitate a basic understanding of the argumentation through a retrospective overview and insights into the author's current research on the topic. Initially, emotional aspects are mentioned only implicitly within the scope of artistic expertise, not determined explicitly. But in the compilation of concrete criteria, emotional aspects come into focus, followed by a discussion on whether aspects of the phenomenon[9] termed 'music and emotion' can be better experienced, interrelated, and imparted within the didactic endeavours of instrumental education by taking the described research results into consideration.

2. Strings Biomechanics – Retrospective and Eichhorn's Legacy

The biomechanics of string instrumental performance with respect to artistic aspects of sound production were studied originally as part of applied physiology, and later as part of music physiology[10] or respectively musician's medicine. Research on biomechanical aspects of various performing arts just recently gained autonomous status due to the foundation of the journal *Arts Biomechanics*. Thus, the main focus of research articles can remain on artistic goals for performance, whereas the medical aspects or those of occupational health organization do not need to constitute the main subject.

An early twentieth-century analysis of bowed string instrumental performance by Friedrich A. Steinhausen, a German physiologist, (s. Steinhausen 1928 [1903]) was the first study on bowed string instrument biomechanics, and the results may be still viewed as fundamental on many essential points. For example, he gave evidence for the characteristic and systematic use of forearm rotation in the bowing of string instruments by means of an optical enlargement and measurement of the motion.

Steinhausen showed that pronation and supination, the inward and outward rotation of the forearm, act as a means of bow pressure regulation, in the style of so-called "double-lever-playing" (ibid. 69); that is, with a two-armed lever moving around a fulcrum somewhere between thumb and index finger (to add pressure), or thumb and little finger, respectively (to reduce the bow pressure). But what he found most important was "the realization that all

[7] An empirical examination of the hypothesis that an instrumental education imparting sound physical and biomechanical knowledge leads in the long term to higher artistic expertise (for example, by means of an expert rating procedure) is yet to be realized. This theoretical discussion is based on the aggregate of previous research results, as well as the studies of Hasselbach et al. 2010 and 2011, wherein a selected characteristic of expertise in violin playing was identified by means of three-dimensional movement analysis and tested positively in terms of its trainability in teenage violin students.

[8] The symposium took place on March 11th-12th 2011 at the Center for Interdisciplinary Research (ZIF) in Bielefeld, Germany.

[9] Determining observations about the constructs ,music' and ,emotion' cannot be made at this point. They can be found for example in the articles of Constanze Rora or Jürgen Vogt in the original publication of Krause and Oberhaus 2012.

[10] The establishment of the special field of music physiology in Germany traces back particularly to the endeavours of Christoph Wagner since the 1970s. Cp. Wagner 2005.

practice aims at economy of forces, relaxation, passivity, short impulse and swinging movement. [...] Only a relatively passive limb can be swung; passivity and swinging movement belong together" (Steinhausen 1928, X).

The physiologist Wilhelm Trendelenburg illustrated, among other effects, the workings of the "artistic slant" ["künstlerischer Schrägstrich"] (Trendelenburg 1974 [1925], 33) – a technique for shifting the contact point[11] by means of setting the angle of the bow in relation to the vibrating string, which will be presently described in more detail. Thus, essential, artistically relevant systematic movements on the coronal plane and transverse plane[12] were described already by these two researchers.

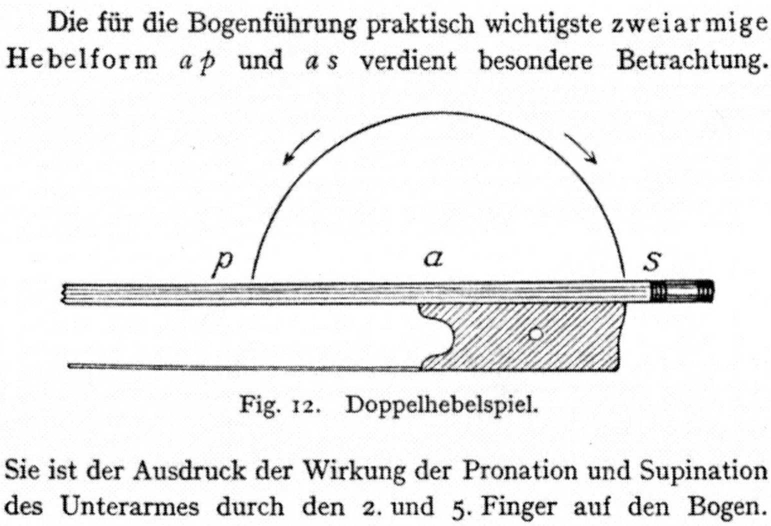

Figure 1. Illustrative example of double-lever-playing, taken from: Steinhausen 1928, 69. Caption to the illustration: "The two-lever profile that is most crucial for bowing action, as shown by *ap* and *as*, is worth special consideration. This profile represents the effects of pronation and supination of the forearm through the second and fifth fingers on the bow."

The significance of motions on the sagittal plane was further demonstrated by August Eichhorn[13]. Whereas Steinhausen had still believed (in the beginning, but not later on, see the above quotation) and published that "up-bowing and down-bowing" (Steinhausen 1928, 44) leads to constant loading of the musculature, and also Hugo Becker[14] had still cautioned against letting the elbow deviate from the level of the (quasi transverse) playing plane, Eichhorn pointed out that the entire arm can move with extreme economy oscillating in form of a two-armed lever on a parasagittal plane – whether in the large movement of a full-length bowing stroke (whole bow), or in small, repetitive-style strokes. Eichhorn's theory of an oscillating leverage system in bowing distinguishes itself by showing how multiple two-

[11] Due to the complexity of the topic, not all of the subject-specific terminology will be explained.

[12] The terms 'transverse plane' (e.g., the floor), 'coronal plane' (e.g., the surface of the forehead), and 'sagittal plane' (e.g., a division passing through the centre line) are used to describe movements in a physically oriented coordination system. Here, an underlying simplified observation model is assumed, in which the playing plane is roughly identical to a transverse plane; for example, the approximate playing position for the D or G string on the violin or on the violoncello.

[13] See Hopfer 1941.

[14] See Becker and Rynar 1971 [1929], 20ff.

armed levers function in an interconnected oscillating network (including Steinhausen's 'double-lever'); while the alignment of the angled position, in terms of the 'artistic slant', remains independent of this oscillating leverage system.

Efficient systems of leverage in bowing were developed by some musicians throughout the 20th century due to constantly increasing standards for artistic achievement, in order to turn their 'objective overload' (as it is termed in work physiology) into a reduced 'subjective load', and thereby, to enable the artists to stand out from the striving masses once again. Examples of these leverage systems could be seen in the playing of Pau (Pablo) Casals (1876-1973), Emanuel Feuermann (1902-1942), and David Oistrach (1908-1974)[15]. After Feuermann's U.S. American debut in 1935, a critic wrote, "Difficulties do not exist for Mr. Feuermann, even difficulties that would give celebrated virtuosi pause", (Morreau 2002, 93).

Eichhorn had studied with Feuermann in 1920 in Cologne. Afterwards, out of scientific and educational interest, in addition to his work as soloist cellist and music academy professor, he analysed Feuermann's playing as a prominent example of high expertise. Eichhorn's research resulted in a very idealistic teaching concept, which can be characterised as follows: All recommendations on 'technique' in bowed string instrumental performance should be made only on the basis of scientifically documented findings. Furthermore, priority should be given to the learner's ability to be self-determining within the structured framework of their person, the music and the instrument, in order to avoid negative effects from questionable subjective theories from the teacher). The goal is to enable a reliable, conscious production of sound that will be enduringly accessible, even under the high demands of an orchestra position or a solo career, and which supports self-initiated creative development that is, by and large, free from the occurrence of coincidental sound results.[16]

Eichhorn's theory of a mass-balancing oscillating leverage system in bowing is thus far unique in its complexity and its scientific validity, since it builds strictly upon research findings from biomechanics and sound physics.[17] The scientific validity of Eichhorn's theory was first empirically established in a 2010 study of expertise by Julia von Hasselbach, Wilfried Gruhn and Albert Gollhofer, which examined the occurrence of mass-balancing oscillations at various levels of expertise.[18] A comparison of bowing by students of musical education (future school music teachers) with that of students of instrumental pedagogy (future violin teachers), as well as with that of students in a purely musical course of study (future solo performers), showed that the manifestation of mass-balancing oscillations differed significantly among levels of expertise and correlated strongly with the respective average practice times per day (as calculated over the entire course of study in violin). Mass-balancing oscillations in the bowing arm can thus be viewed as a characteristic feature of high expertise in violin playing.

---

[15] To get an impression of the economy and differentiation possible in cello playing while using oscillating leverage systems, one might consult film documents like the 1954 recording of Pau (Pablo) Casals' playing, or the 1939 recording of Feuermann's playing. Even though the motions of these musicians in the recordings have not yet been empirically quantified by video analysis, to the practiced eye, the oscillation of the system and the resulting impression of ‚unfastened playing' can be recognised.

[16] According to the deliverance of Eichhorn's former students such as Josef Schwab and Gerhard Mantel; see also Clemen 1999/2000, 30.

[17] See Palac 1992 and Hasselbach 2009.

[18] See Hasselbach et al. 2010.

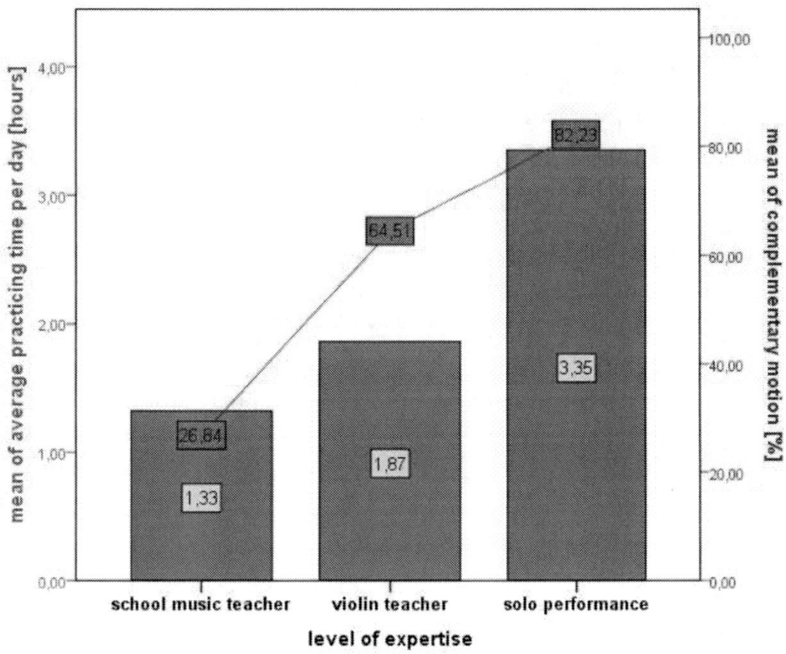

Figure 2. Graph of the results of the study on expertise, in Hasselbach et al. 2010, 392. Columns: average practice times per day in hours. Curve: average complementary motions of the hand and elbow, expressed as percentages (i.e., mass-balancing oscillations of a central lever in the Eichhorn leverage system).

A central characteristic of an oscillating lever technique is a complementary motion of the hand and elbow around a fulcrum (in the forearm, near the elbow joint). When a very low percentage of complementary motion is present, an extensive parallel motion appears in the hand and elbow; this indicates a comparatively static effort and a constant strain on the musculature.

With a high percentage of complementary motion, inertia becomes the propulsion, as in a mobile, for a flowing bowing motion that is free from acceleration after the initial impulse; and the muscle activity needed is reduced, so that relatively small impulses suffice as catalysts for motion.

The flow of bowing motion, which may at first glance appear 'transverse', is thus superimposed by a 'sagittal' rocking of the bowing arm, as in a seesaw. The regulation of bowing pressure is also thereby equally supported, in the upstrokes as well as the down strokes, by the transformation of stored energy into motion (that is, from potential energy into kinetic energy). Simultaneously, the mass of the bow, as long as it is not completely involved **in a musical task, is also fundamentally 'levered' in balance, so that the otherwise variable effect of the weight of the bow on the string is 'neutralised', and the bow exists from the frog** to the tip in a state of constant readiness.

Thus 'hair-triggered'[19], the most delicate of impulses (especially through the rolling of the forearm) from emotional states can fire directly into the system of play, creating an

[19] **'Hair-trigger' ('Matchabzug' in German), in firearms** terminology means a trigger that is so finely pre-cocked, that the slightest motion of a finger fires the shot, while the otherwise greater muscular effort required would compromise the precision of the aim.

audible effect with hardly any loss of impact, as long as all the other parameters of sound physics are in alignment with each other. The fundamental ability to consciously learn and to train these mass-balancing oscillations in violin bowing has been empirically researched and confirmed among violin students aged 10 to 15.[20] While one training group increased its percentile of mass-balancing oscillations significantly, from 31% to 56% after fifteen weeks of additional trainings, a control group taking normal violin lessons in the same time span altered its values insignificantly, from 23% to merely 28%.

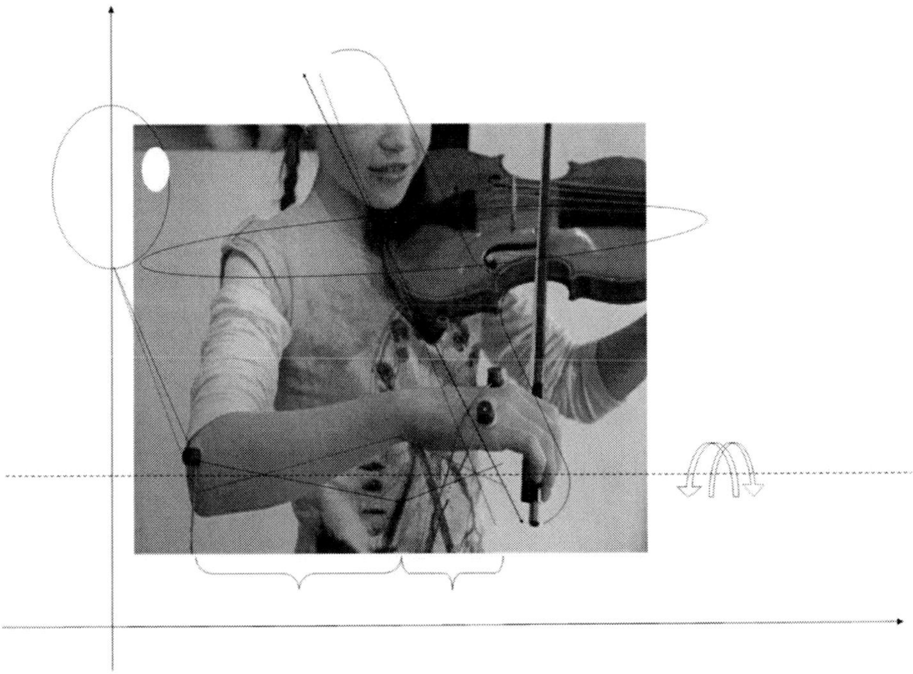

Figure 3. A schematic approximation of Eichhorn's theory of a leverage system of interconnected oscillating two-armed levers (indicated within the brackets around the arm and hand, as well as at the bow. In addition: the action of Steinhausen's doublelever with the rolling rotation of the forearm, the arrows indicating pronation and supination). Three-dimensional movement analysis enabled by three ultrasound transmitters (equipment brand: Zebris CMS 20) on the hand and elbow, in the study by Hasselbach et al. 2011.

To achieve artistically highly expressive and, as the case may be, emotionally effective results, the biomechanical possibilities and the requirements of sound physics definitely must be ideally attuned to one another. In Figure 5 the coloured field represents the area of relation between the point of contact and bowing pressure (with constant bowing speed as a given), in which the sound of a bowed string instrument can be differentiated in its harmonics without losing the fundamental tone. The closer to the bridge one plays, the harder it becomes to manoeuvre the bow within the narrowing coloured field, although the sound should remain musically differentiated.

[20] See Hasselbach et al. 2011.

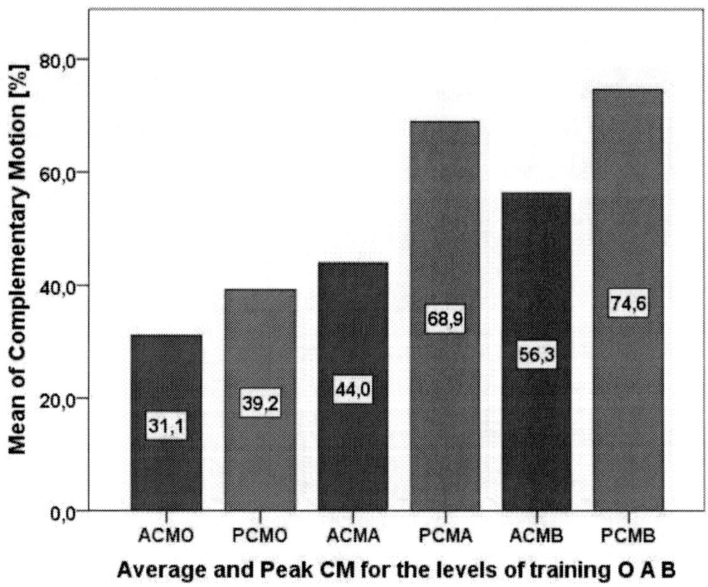

Figure 4. The mean values of the groups' average levels as well as of the peak levels of complementary motion achieved at the training points in time O (start date), A (after 5 weeks of training), and B (after fifteen weeks of training), in the study by Hasselbach et al. 2011.

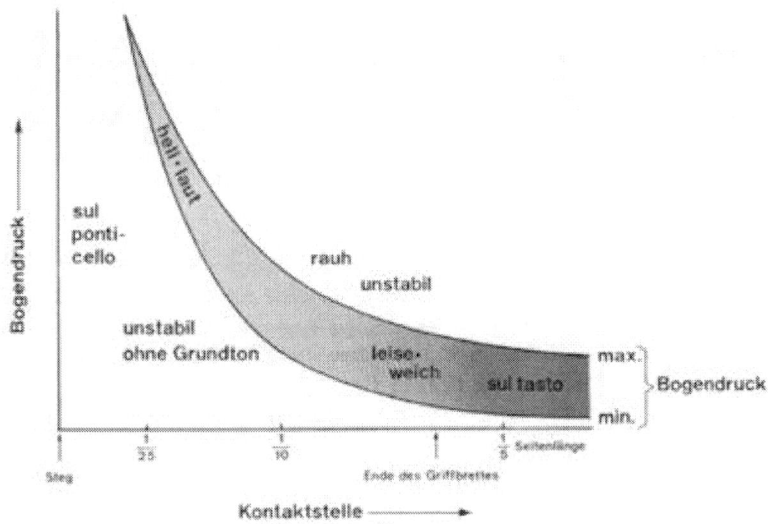

Figure 5. The dependent relationship between bowing pressure (y-axis) and point of contact (x-axis) under a constant bowing speed. Taken from Meyer 1978: Illustration of the research results of J. C. Schnelleng.

A great number of the demands on soloist playing take place in the tip region of this field, requiring a high reliability in bowing motion and bowing pressure. It is easy to imagine how much more difficult it becomes to hit this region, when it is wavering up and down due to unintended and uncontrolled fluctuations in bowing speed. A fundamental division of the biomechanical function to produce bowing speed from the function for differentiating bowing pressure, into two relatively independent biomechanical processes, was, therefore, a ‘revolutionary’ development in playing technique for bowed string instruments. This is a further central component of Eichhorn’s theory.

At this point, the already-mentioned “artistic slant” of Trendelenburg, also called “the slanted bow”[original: “schiefer Bogen”][21] by Gerhard Mantel, can be presented in more detail in relation to the sound field of John C. Schelleng[22] explained above. The artistic slant takes advantage of a drifting effect to distribute the forces of the moving bow within a parallelogram, such that one vector makes the thrust of the bow proportional[23] to the chosen degree of the angle, whereas the other vector generates extremely ‘pure’ transversal vibrations of the string.

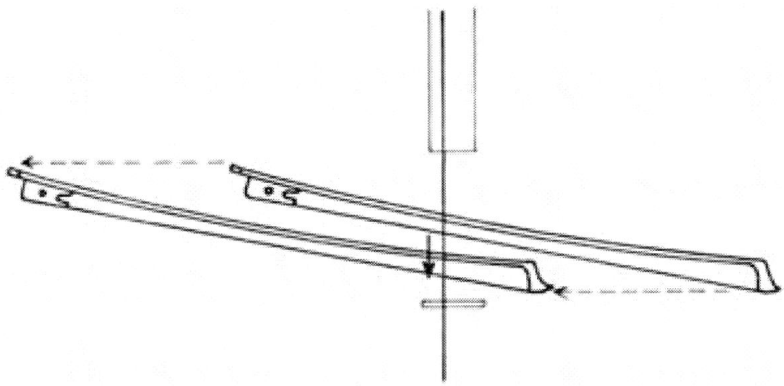

Figure 6. An illustration of bowing drift (small arrow pointing toward the bridge) in cello playing, with the bow at a ‘slanted’ angle to the string, but with the bowing motion at a right angle to the string (broken-line arrows). Taken from: Mantel 1999 [1987], 61.

This leads to a resulting sound without incidental unintended noises that arise from a haphazard thrusting motion through its accompanying longitudinal vibrations due to friction, which, though it may certainly be intentional in some contexts of 20th century classical or contemporary music, is generally still viewed as a characteristic sign of a lack of expertise. In the “student’s slant” [original: “schülerhafter Schrägstrich”] (Trendelenburg 1974 [1925], 33), the bow is held tightly with greater pressure against the forces of drifting (not far from the ‘crushing point’, i.e. the upper line of Schelleng’s sound field), by which the sound then becomes not only accompanied by this constant friction noise, but also monotone in its colour. Yet brilliant colours, and a sound with good carrying capacity, arise from a bowing pressure so minimal that it is almost to the point of ‘flautato’[24] (near the lower line of Schelleng’s sound field) and require a finely developed sensitivity for the occurrences of drifting. An artistic shaping control over the degree of the angle, and a corresponding proportional adjustment of the bowing pressure, enable the musician to perform ‘organic’-

[21] See Mantel 1999 [1987], 59ff; Mantel 1976, 166.
[22] See Meyer 1978.
[23] The more the bow deviates from a right angle in relation to the string, the more, and/or faster, the point of contact skids toward the bridge or toward the fingerboard, its direction depending on the degree of deviation from the right angle.
[24] See Hopfer 1941; Meyer 1978; and Mantel 1999 [1987].

sounding tone relationships, with dynamic variations during tone progression, as well as the selection of appropriate sound colours, without hindering the speed of the flow of bowing.[25]

It would overreach the bounds of this article to try to make the status of these findings understandable in detail. However, Eichhorn's theory of a mass-balancing oscillating leverage system in bowing lays the foundation for the following roughly outlined remarks. These will present criteria for performance that not only optimally meets the sound physical requirements of bowed string instruments, but uses the biomechanical potential of human beings very economically to generate sound configurations that are differentiated and possibly perceivable as emotionally expressive.

3. A-Marks and B-Marks: Criteria for Full Points in Creating Sound

In order to produce an even current of sound with good carrying capacity, as Röbke describes it with his observed violinist, a number of criteria must already be fulfilled. In the following remarks, similarly to the A-mark ranking for figure skating, 'Criterion A' will be used for a clean technical execution; whereas 'Criterion B' will apply to those further cases that achieve a differentiated modulation of this current of sound, in the sense of an artistic performance.

A more powerfully shaped creation can potentially evoke a more intense aesthetic concurrence in the listeners and, given certain conditions within their personal neural 'circuitry', can become further associated with emotions.

The Progression of Bowing Pressure

One element of making sound within Criteria A is a proportional increase in pressure, in the process of causing the string to vibrate from a silent position to the intended amplitude. This produces the least possible articulation.

After the proportional increase in pressure in the process of building vibration, which involves only a few undulations, an immediate reduction of bowing pressure to a substantially lower level is necessary to keep the achieved amplitude of vibration going, for otherwise the vibration will soon be disturbed by an unchanged bowing pressure exceeding the necessary level at the chosen bowing speed, leading automatically to a timbre near the 'crushing point' of the sound field.

The needed pressure progression can be best achieved via a slight ballistic effect, because the time frame for creating the amplitude of a vibration comprises only a few milliseconds. Gross-motor activity would require too much time to deliver a rhythmically precise monotone sound without stereotypical disturbances.

---

[25] Without moving the point of contact, dynamic variation must be realised by altering the bowing speed, which can succeed only in connection with proportionately appropriate changes in bowing pressure. These changes often can be carried out only in the course of audible articulations that are, however, not always intentional, especially not during a continuous, dynamically supported increase or decrease of tension.

In a mass-balancing oscillating leverage system, the creation of sound happens almost 'on its own', with as little muscular control as possible, but with just the necessary amount of balancing adjustment in the kinetic chain.

This leverage system, in which the masses of the bow and the arm work together, delivers the constant effective bowing pressure required for an evenly formed pattern of sound, wherein only the differentiated fine-tuning is carried out via the muscle tone of the force-conveying bowing hand.

Finally, the change of direction in the oscillation of the armlever also provides the necessary minimal release of bowing pressure at the moment of the shift in bowing direction, and gives a nearly-direct renewed proportional increase in pressure; thus, the bow is able to almost inaudibly catch the string while still vibrating unimpeded at its midsection, and to augment the amplitude again on the reversed direction of the stroke.

Rhythmic Bowing Speed

A further criterion for an evenly formed current of sound is a rhythmic bowing speed. In 'resonance' with the circulating right-angled vibrations of the bowed string, this bowing continually catches the string in the rhythm of its circuit, since otherwise the 'freedom' of the vibrations would be limited. As long as the point of contact and the amplitude always remain the same, the rhythmic bowing speed is also constant. But with a wandering point of contact, the bowing speed must be adjusted in relation to it, as long as the change in the point of contact's position is not also connected to a change in dynamic through correspondingly matched bowing pressure; whereas the amplitude at the wandering point of contact would be identical at every second.[26]

Transverse Oscillations

The criterion for a mostly 'scrape-free' sound production, beyond an appropriate progression of pressure and a rhythmic bowing speed, is the creation of transverse vibrations on the strings: Either through a steady bowing motion in an almost right-angle proportion to the string, and/or through the mastery of the 'artistic slant' already illustrated, that is, through the function of letting the bow drift on the vibrating string. The ability to adjust the angle can be trained and optimised through hearing and proprioception, and partly visually. But drifting can take place only under the precondition of a free bowing flow of 'low-pressure' playing, that is, without muscular restraint against the drift.

When a 'perfect' expression of sound, according to Criteria A, is criticised for its absence of "emotional message" (Röbke 2008, 9ff), then the following basic possibilities exist: 1. The violinist could, but doesn't want to. 2. The violinist wants to, but cannot. In the former case it would be no problem to correct by imparting knowledge from sound physics and biomechanics. Yet in the latter case the problem may be resolved, as long as it is a matter of a

[26] In this way, for example, when moving the point of contact toward the bridge while the dynamic remains stable, the bowing speed could be reduced in order to ‚save' bowing; or, with a constant bowing speed, the bow could drift toward the fingerboard, in order to achieve a decrescendo without losing the rhythmic bowing tempo.

pure instrumental-technical incompetence, and the violinist possesses a sufficiently strong 'feeling' in the sense of energetic-creative intentions which could be transformed into music, if the violinist only knew how it might be done. In contrast to a sound of 'static' perfection, a differentiated, 'dynamic' creation of sound, such as that which Röbke found lacking, can be achieved through the following Criteria B:

Dynamic Change of the Point of Contact

The ability to control drifting through anterior and posterior arm movements[27], i.e. the adjustment of the bowing angle to the strings, is used to consciously change the point of contact during the duration of a note or a series of notes, so that, while holding to the criterion for pure transversal oscillations, flowing dynamic developments in tones can be shaped.

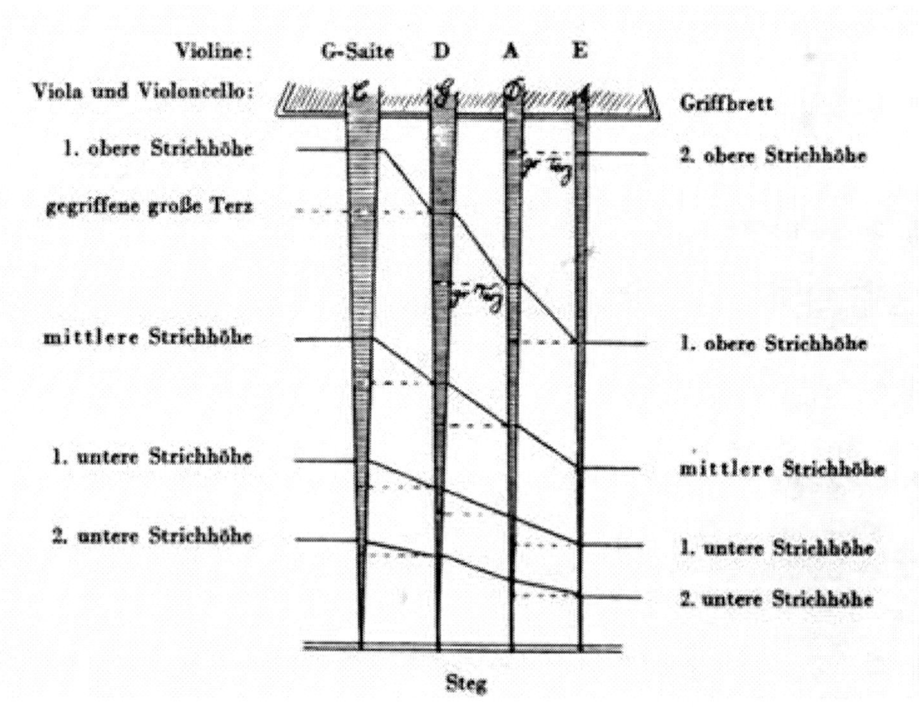

Figure 7. Proportional changes of the points of contact (heights of stroking) needed to obtain the equivalent sound when changing strings. In Hopfer 1941, 12.

Internal build-ups of tension thereby receive an audible equivalent with the greatest possible spectrum and the highest sound quality. In any case (along with an appropriate rhythmic bowing speed), in addition to the 'normal' progression in bowing pressure, there must also be a proportional adjustment of bowing pressure to the changing point of contact.

Dysfunctions in the development of sound may result especially from a failure to consider that, while shaping tone relationships across changes of strings, the equivalent points

[27] Especially in the case of violin or viola playing the movement really goes forth and back, whereas in violoncello or double-bass playing the movement may be felt in the same way in relation to the body, but in fact it is more like a down and up going movement, because of the slanted position of the instrument.

of contact (the heights of stroking) have a proportionally changed position on strings of varying widths. A static retention of the same distances from the bridge can thus lead to complications within the development of sound, to unintended ruptures in sound intensity (dynamics) and changes of timbre, if not compensated by a corresponding adjustment of other parameters. Yet an over-involvement in 'compensatory activity' can impede the actually intended shaping, wherein the outcomes are often makeshift solutions or seemingly nearly aimless developments. Similar proportional shifts of the equivalent points of contact may also appear on a single string, through the act of fingering and thereby shortening the vibrating string when playing in higher positions.

Independence of Bowing Pressure Regulation Functions

When adjusting the bowing pressure, there may be great variations in the resulting sound, depending upon how this parameter is regulated. A very symptomatic difference appears between a rigid, static and mostly over-exaggerated 'pressure' via the 'index finger-thumb-lever' (the pronation side, in Steinhausen's double-lever model) and a 'settling' of 'the weight of the arm' through the middle fingers onto the bow stick. While static finger pressure creates muscular tension progressing from distal to proximal in the kinematic chain, 'resting' weight offers relaxation and 'permeability' to oscillations in the kinematic chain. By applying a technique that uses the oscillating weight of the arm for greater effects, the finest tensions, in the sense of 'delicate inner sensitivities' or 'shimmering colours' may be modulated in as well, via the still freely functioning forearm-rolling action of the bowing arm, because this part is not statically locked into the task of producing the basic bowing pressure, but stays flexible for oscillating motions.

Finally, it is possible to recreate alternating energetic states analogous to human states of excitement, mainly through the varied effects of 'built-up' and 'released' bowing; i.e. through different sound colours based in the interplay of the triad of bowing pressure, bowing speed and point of contact, in connection with intensities in dynamics, vibrato and articulation. The prerequisite for this, however, is again a mastery of drifting in the balancing act of keeping a freely flowing bow at the position of the point of contact required with respect to the intended area of the sound field. If the point of contact is 'forced' through increased bow pressure, then the brilliant and 'sound-carrying' regions of the colour palette, as found in unfettered playing within a strong *forte* or an intensely 'singing' *piano*, cannot be created, and these effects, that are especially difficult to master, remain - regardless of individual ideals for sound – outside the realm of possibility.

Acceleration-Free Articulation

At the same time, accentuations or articulations should be able to be layered without disturbance over the dynamic change of the point of contact. This takes place through impulses reaching into the depth of the strings, originating from the rolling action of the forearm, as well as from the mass-driving effect of the bowing arm. Attempts to accentuate or articulate by accelerating the bowing speed without a corresponding 'in-depth-dynamisation' of the string (acceleration of the vibration to a greater amplitude) produce an undesirable stridency of colour; so that often, due to a lack of useable sound results, clear accents and articulations are unconsciously foregone. Yet they may emerge especially well and 'risk-free', if the bowing arm can balance its mass in the motions of an oscillating leverage system. Neglecting these dynamic movement components that operate in the depth of the strings is a cause of (in the truest sense of the word) superficial-sounding, static production of sound in bowed string instrumental performance.

Vibrato with a Rotational Component and Interplay of Colours

Rather than producing vibrato primarily through flexion and extension in the joints of the grasping arm, it can be created in cooperation with the rolling of the forearm (rotation). In this way, vibrato may be generated from a swelling of oscillations in the leverage system there, arising with building 'emotions'. In Eichhorn's view, the leverage system of the grasping arm differs only functionally, but not structurally, from the leverage system of the bowing arm. A 'back and forth' movement (flexion and extension) of the arm produced with greater muscular force is generally accompanied by a stiffening of the vibrating finger under an increased pressure on the fingerboard. The increased pressure becomes necessary in the face of the 'attacking' forces to avoid a slippage of the tone pitch. As a result, monotone changes happen in frequency and amplitude. The rotation of the forearm is a frequently neglected component of vibrato. Yet it has the advantage that, due to its rotation around a physiological central axis, hardly any stabilising opposing forces must be expended, and the finger joints may stay actively flexible to animate the oscillations of the leverage system. On the condition that the proximal joints, in particular the shoulder, are not blocked from oscillating by inhibiting muscle activities, a feeling-like impression of, for example, 'cold' or 'hot', respective to the intensity of the vibrato, can then become thereby differentiated and intuitively influenced. But only a vibrato that is concurrently supported by iridescent coloration via dynamically oscillating forearm rolling of the bowing arm, that is, by a variation in bowing pressure, is the sort that listeners tend to hear as high-value quality or 'touching'.

It is only by applying these sound-differentiating techniques, that an impression of 'liveliness' and 'richness of relationships' in the occurrence of sound may first arise – even though, in the exuberance of a performance situation, that which was trained in the process of practice needs no longer be conscious. Mastery of technique may help to carve out musical structures and, in the sense of a process of melding music and emotion together, to support that which Röbke calls "a glowing saturation of the tonal material" [original: "Durchglühen des Tonstoffs"] (Röbke 2008, 11).

Conclusion

The extensive explanation of biomechanical and sound physical aspects in bowed string instrumental performance shows that, for an effective expression or for the impression of emotion in music, satisfying the basic Criteria A is not sufficient and playing of this kind cannot be described as 'perfect technique'. This requires the further comprehensive fulfilment of the Criteria B, which makes possible the finest and strongest differentiations of the musical parameters. When a skills training to fulfil the Criteria A is attached to side effects like **'becoming** hollow' **or** 'loss of inspiration' as Röbke suspects, then this sort of training deserves extreme criticism. More preferable would be a form of instruction that encompassed both A and B Criteria from the very beginning, taught with awareness of current scientific findings.

It may be assumed of instrumental pedagogical training that a tendency to concentrate on Criteria A is prevalent. The reasons for that lie in either a lack of knowledge or a lack of understanding of Criteria B. The sound physical-biomechanical foundations of the creative aspects of A and B should, however, be a part of the methods training of all students, whether in the artistic or the educational courses of study in music universities. These foundations should be conveyed in theoretical and practical ways, in which theory, perception, practice, transfer, and stabilisation are central consecutive steps. In the first place, this would foster long-term accessibility to creative possibilities in daily professional practice and the sustainable health of the physical apparatus of motion. Furthermore, the comprehensive training of instrumental instructors would already lay the groundwork within music schools and in private instrumental lessons with children, for a successful association of music with emotion. For students, engaging with current scientific findings on the playing of their own instruments, as well as being confronted with diverse didactic viewpoints that are independent of the authority of their professors or teachers, should support flexibility of thought, and have long-term positive effects upon an autonomous instrumental-technical and artistic development; particularly upon the ability to let their own sensitivities and intentions become audible through the instrument.

Would it not be wonderful if, in a circular form of cognitive-emotional development, the skills for sound differentiations on an instrument could give wing to a differentiated **development of the musical 'emotional life', and vice versa? A contrasting negative** image would be that the existing or pedagogical-psychological supported emotions had to constantly struggle in a behaviouralist manner against a wall of technical obstacles, until an escape hatch appears, through which emotion may become audible. An escape hatch that might, with further attempts, be widened, yet sometimes also closes again, or simply stays shut for long periods of time. However: acting, or respectively, bowing consciously to bring musical structures to the sound, and empathising with the associated emotion, and thereby learning to recognise the feeling with the aim of becoming able to intuitively put emotion into tones as needed while making music, may possibly offer a promising way to success.

REFERENCES

Becker, H. and Rynar, D. (1929/1971). *Mechanik und Ästhetik des Violoncellspiels.* Wien: Universal-Edition.

Clemen, J. (1999/2000). *Ein Orchester aus vierzig Celli. Symposium zum 100.* Geburtstag August Eichhorns. Neue Musikzeitung, 12, 30.

Röbke, P. (2008). Einleitung. In Figdor, H. and Röbke, P., Das Musizieren und die Gefühle. *Instrumentalpädagogik und Psychoanalyse im Dialog,* Mainz: Schott, 9-30.

Hasselbach, J. von (2009). 100 Jahre Physiologic Turn in der Streichinstrumentalpädagogik. Eine Bestandsaufnahme. In Schläbitz, N. (Ed.), *Interdisziplinarität als Herausforderung musikpädagogischer Forschung*. Essen: Blaue Eule, 239-262.

Hasselbach, J. von (2012). Für ‚lebendige' und ‚beziehungsfähige' Töne im Streichinstrumentspiel. Eine Reaktion auf Peter Röbkes Einleitung zu Das Musizieren und die Gefühle. In Krause, M. und Oberhaus, L. (Eds.), Musik *und Gefühl. Interdisziplinäre Annäherungen in musikpädagogischer Perspektive.* Hildesheim: Olms, 67-88.

Hasselbach, J. von and Gruhn, W. and Gollhofer, A. (2010). Mass balancing oscillations. An indication of expertise in the bowing of violinists. A quantitative micromotion study. In Demorest, S.M. and Morrison, S.J. and Campbell, P.S. (Eds.*), Proceedings of the 11th International Conference on Music Perception and Cognition* (ICMPC 11). Seattle: University of Washington, 388-393.

Hasselbach, J. von and Gruhn, W. and Gollhofer, A. (2011). Effects of training on mass balancing oscillations in the bowing of (pre) teen violin students. A quantitative micromotion study. *Arts BioMechanics,* 1, 1. New York: novapublishers, 1-14.

Hopfer, M. (1941). Die Klanggestaltung auf Streichinstrumenten. Das Naturgesetz der Tonansprache. Kurze Einführung in „Die gestaltende Dynamik der Bogenmechanik" von August Eichhorn. Leipzig: Kistner and Siegel.

Mantel, G. (1972). *Cello Technik.* Bewegungsprinzipien und Bewegungsformen. Köln: Gerig.

Mantel, G. (1975/1995). *Cello Technique. Principles and Forms of Movement.* Translated by B. Haimberger Thiem, Bloomington: Indiana University Press.

Mantel, G. (1987/1999). *Cello üben.* Mainz: Schott.

Morreau, A. (2002). Emanuel Feuermann. New Haven/Connecticut: Yale University Press.

Meyer, J. (1978). Physikalische Aspekte des Geigenspiels. Ein Beitrag zur modernen Spieltechnik u. Klanggestaltung für Berufsmusiker, Amateure u. Schüler. Siegburg: Schmitt.

Palac, J. A. (1992). Violin Bowing Technique. An Analysis of Contemporary Pedagogical Literature According to Principles of Human Movement. *Medical Problems of Performing Artists*, 7, 30-34.

Steinhausen, F. A. and Reuter, F. von (1903/1928). Die Physiologie der Bogenführung auf den Streich-Instrumenten. 5. edition. Leipzig: Breitkopf and Härtel.

Trendelenburg, W. (1925/1974). *Die natürlichen Grundlagen der Kunst des Streichinstrumentspiels.* Unchanged photomechanical reprint of the edition 1925. Kassel: Hamecher.

Wagner, Ch. (2005). *Hand und Instrument: Musikphysiologische Grundlagen.* Praktische Konsequenzen. Wiesbaden: Breitkopf and Härtel.

In: Applications in Arts Biomechanics
Editors: Gongbing Shan and Peter Visentin
ISBN: 978-1-62808-015-5

Chapter 2

THE DEVELOPMENT OF MOTOR COORDINATION AND MUSICAL ABILITIES IN PRE-SCHOOL CHILDREN

Wilfried Gruhn*[*,1], Madeleine Haussmann[2], Ulrike Herb[2], Cornelia Minkner[2], Katrin Röttger[2] and Albert Gollhofer[2]

[1]University of Music, Freiburg, Germany
[2]Institute of Sport and Sport Science, University Freiburg, Germany

ABSTRACT

Movement and musical abilities develop simultaneously. This study investigates the correlation between musical and motor development. Observational data were collected from musical, cognitive and motor tasks. Twenty-eight children, aged 3.6- to 6.6- years-old, performed a standardized motor test measuring fine motor abilities, balance, motor reaction, and motion control, as well as a music aptitude test and three nonverbal subtests of a cognitive assessment test. Motor abilities and musical aptitude as dependent variables were related to gender, age, and cognitive state as independent variables. The results revealed a significant correlation between motor and musical abilities, as well as a linear progression. A second experiment introduced biomechanical and neurophysiological data from a subset of the sample and related these data to the measures of music aptitude. In general, these findings support the results of the first experiment.

Keywords: Motor coordination, motor control, balance, postural stability, music aptitude, electromyography, proprioceptive amplification ratio

---

[*] Corresponding author: Wilfried Gruhn, Lärchenstr. 5, 79256 Buchenbach, Germany; E-mail@wgruhn.de; phone: +49-7661-1624; fax. +49-7661-99653.

INTRODUCTION

It is a common everyday experience that people often move to music. Moreover, music is typically more likely to evoke body movements than are other expressive modalities, such as speech. A behavioral study reported motor behaviors stimulated by listening to either music or speech in two groups of infants aged 5 – 24 months (Zentner and Eerola, 2010). The children engaged in significantly more rhythmic movement to music and other metrically regular stimuli compared to speech. This can be interpreted as an early and quasi-"natural" connection between music and movement that becomes strikingly obvious in dance-like activities performed in children at play. Later on in life musical activities are often driven by the intent to develop excellence in high performance. In such situations movement is mainly related to the physiological conditions of finger, hand and arm movements in instrumental practice. The focus on training for technical achievement in music, then, parallels instrumental practice with competitive sports.

Until now, little attention has been paid to the general structure of an interaction between motor development and musical abilities in childhood. In music learning theory, body motion has become a major indicator of musical abilities in general (Gordon, 2001). Therefore, the relation of body movement and music learning has been stressed in music education (Danuser-Zogg, 2002; Gruhn, 2010; Hodges, 2009; Malinowski, 2000). This can be traced back to the ideas of Émile Jaques-Dalcroze (Jaques-Dalcroze, 1977) or Moshe Feldenkrais (Feldenkrais, 1976). In recent times, the awareness of body consciousness and music (Shusterman, 2008) appears in the broader context of the philosophical dimensions of music perception and cognition, and has introduced the concept of somaesthetics (Bowman, 2010).

In a developmental context, a study on neonates underpins the evidence for the presence of a sense of pulsation in humans that is related to a regular motion sequence. The perception of a regular pulse in an auditory signal might be innate or learned by very early exposure to the mother's heartbeat (Winkler, Háden, Ladinig, Sziller, and Honing, 2009). Newborn infants who listened to a rhythm that kept a regular beat quickly developed an expectation for the onset of a new metric cycle, even when it was not marked by stress. The omission of the downbeat caused a clear ERP signal associated with violation of an expected regular continuation. A regular pulse which determines an underlying rhythmic structure is fundamental to human perception and underpins how rhythmic regularity and perception are related.

A strong relationship between music and movement becomes evident in the ability to synchronize movements with an external sound stimulus. Young children are able to synchronize their body movements to a musical pulse, i.e., they move their hands or feet in synchrony with a tapped rhythm or a sung tune without having visual contact with the sound source (Thaut 2003; Trevarthen, 1999). This has also been documented in parrots, which can adapt their head nodding to the pulse of the music they hear (Patel, Iversen, Bregman, and Schulz, 2009; Schachner, Brady, Pepperberg, and Hauser, 2009). The ability to entrain movements to an external timekeeper raises the question as to whether there is an endogenous predisposition to connect sound and movement by auditory motor integration.

If we consider the literature regarding the development of coordinated motion (e.g., as reflected by the child's first walking independently) relevant studies show that coordination and proprioceptive motor sensitivity are interrelated (Berger, Altenmüller, and Dietz, 1984;

Chang, Kubo, Buzzi, and Ulrich, 2006; Kubo and Ulrich, 2006; Thelen and Cooke, 1987). It seems reasonable and obvious that especially music has a considerable impact on the development of fine motor control caused by instrumental practice which, in turn, also impacts aural differentiation, melodic and rhythmic accuracy, and metric stability. Hence, observational studies have demonstrated that children who can sing more properly in tune and keep a regular steady beat also exhibit well-coordinated body control and the ability to move in space more smoothly in a continuous sustained flow (Gruhn, 1999).

All of these studies support the efficiency of a strong auditory-motor link which is prevalent in vocal learning. This was first demonstrated by research on birdsong (Jarvis, 2004; Marler, 2000; Zeigler and Marler, 2004). A unique neural mechanism connects auditory and motor processing in songbirds, some cetaceans, and humans. Physiologically, a neural connection is established at a very early stage of the central auditory pathway of sound propagation, presumably in the inferior colliculus, where sensory input and motor stimulation are connected. With respect to young children little research has focused on the first appearance of auditory-motor interaction, and it is still unclear whether there is an endogenous disposition for an auditory-motor interaction. This link might be found in the vestibular system, which is located in the inner ear and is responsible for the sensation of balance and motion.

In general, research has shown that body movement plays a crucial role in the learning process. Children do not acquire abstract knowledge, but concrete experiences are holistically acquired by the entire body.

It has been demonstrated that children prefer rhythms to which they were bounced for some time (Phillips-Silver and Trainor, 2007, 2008). The body and its movements generate those proprioceptive signals that are equally important for aural perception and discrimination. In the light of these findings it seems reasonable to investigate the proprioceptive and neuro-physiological mechanisms of a potential correlation between music and body movement that is evident in early childhood, and that may determine the synchronous development of motor and auditory abilities.

2. Experiment I

2.1. Aim of the Study

Given the scientific evidence of an auditory-motor link that is especially effective in vocal learning, and from the importance of body movement in all children's musical activities, the present study sought empirical data to determine parallels in the development of motor and musical abilities in young children. The main hypothesis is that body control and motor coordination is more pronounced in children who exhibit higher scores in musical aptitude tests.

Accordingly, more details and clearly defined tasks are needed to investigate this connection. Information regarding motor and musical abilities was drawn from a standardized motor test and a musical aptitude test. These data were complemented by teacher ratings of the children's tonal and rhythm abilities. Finally, selected nonverbal subtests of an assessment test provided a measure of the children's cognitive development. The authors suggest that examining the mechanisms that impact on musical and motor development will enable a better understanding of children's mental growth and musical learning.

2.2. Participants and Measures

Twenty-eight German-speaking children (5 male, 23 female; aged 3.6 to 6.6 years, median age 4.9 years) from an early childhood music class participated in the study. All children participated in an Early Music Learning Program based on Edwin Gordon's Music Learning Theory (Gordon, 1997).

All children were of good mental and physical health, and did not exhibit any motor or auditory impairment. Four evaluation measures were used:

1. *Primary Measures of Music Audiation* (Gordon, 1979). The PMMA is an aural discrimination test that measures musical aptitude. Forty pairs of short melodies and rhythms are presented, and the children are asked to compare the patterns and decide whether the second is the same or different from the first.
2. *Motoriktest für vier- bis sechsjährige Kinder* (Zimmer and Volkamer, 1984). The MOT 4-6 is standardized for pre-school children with a test-retest reliability of .97. Eighteen test items are divided into four major performance areas evaluating motor coordination (e.g., jumping rope), fine motor control (e.g., collecting matches into a matchbox with two hands simultaneously), balance (e.g., standing on one leg), reaction time (e.g., catching a falling stick), and action speed (e.g., carrying balls from one box into a distant other box as quickly as possible). The data provide a measure of overall motor abilities.
3. *Kaufman Assessment Battery for Children* (Kaufman and Kaufman, 2007). Three age-appropriate non-verbal subtests of the K-ABC were administered: Gestalt Recognition (pictured objects which are not completely visible), Triangle Reconstruction (replicating triangles from patterns consisting of several two-colored elements), and Digit-span (replicating an increasing number of hand gestures combining fist, angle, or palm gestures). The scores reflect children's cognitive development.
4. *Music Performance Scale*. Three performance abilities (singing of tonal patterns, chanting of rhythm patterns, and movement) were rated by the music teacher during early childhood classes prior to the experiment to provide a measure of children's performance skills.

2.3. Procedure

All parents signed a consent form and completed a questionnaire providing information about socio-graphic data and children's preferred leisure activities (such as sports, choir, or instrumental lessons) as well as hobbies (preferred games and entertainment such as watching television, solving puzzles, viewing picture books, romping, or climbing).

All children were tested individually in a large gymnastic hall, although they arrived in small groups (mostly three at a time).The intention of this arrangement was to increase the children's motivation and stimulate their collaborative spirit. The testing was presented as an entertaining game and was accomplished by providing small incentives (small tokens of toys) after each section (motor test, music aptitude test, intelligence test).

The tonal subtest of the *Primary Measures of Music Audiation* (PMMA) was administered first, followed by the eighteen tasks of the *Motoriktest für vier- bis sechsjährige Kinder* (MOT 4-6). The individual tasks were presented as a course of different "games" to be played. Then, an experimenter administered the three nonverbal subtests of the *Kaufman Assessment Battery for Children*. Finally, all children completed the rhythm subtest of the PMMA following the same procedure as for the tonal subtest.

Children were permitted to rest, move around, or play in the hall as they pleased between all test batteries. Raw data from both the motor and music aptitude tests were standardized according to peer norms, and were transformed into a motor quotient for the MOT 4-6 and a percentile rank for the PMMA.

Since all children had comparable socioeconomic status and similar musical experiences from their early childhood music program, a split-half method was applied based on their music aptitude percentile ranks, with children scoring lower than 50% in one group and children scoring 50% or higher in the other group. The MOT 4-6 and K-ABC scores of both groups were then compared. Correlations between all measures were also calculated.

2.4. Results

The music aptitude, motor, and cognitive scores of the two groups were significantly different. Table 1 shows the data for the PMMA and the K-ABC as a function of music aptitude. The two groups differed significantly on all subtests, including the tonal and rhythm subtests of the PMMA. They also revealed a significant difference in their motor skills. Figure 1 shows the rhythm, tonal, and total PMMA scores as a function of high versus low motor skills.

Similarly, significant correlations were found for all motor-test components (i.e., coordination, fine motor skills, balance, and motor control) and PMMA music aptitude scores (Table 2). The total mean scores for motor ability and music aptitude showed a clear linear progression (Figure 2). Subjects who exhibited higher motor scores also had higher percentile ranks in music aptitude, and vice versa (r = .579, $p < .05$, two-tailed). The higher the motor scores (as reflected by the motor quotients) the higher the percentile ranks for tonal, rhythm and total PMMA scores. One could argue that these findings simply reflect an age effect. In fact, a one-way within-subjects analysis of variance (ANOVA) with age as a factor and music aptitude, motor, and cognitive scores as dependent variables found a main effect of age ($p <$ .01). However, a partial correlation with age as a control variable still demonstrated significant correlations for PMMA and the motor quotient (R = 417, p = .034) and, more interestingly, it showed a significant correlation for rhythm scores and fine motor abilities reflected by the motor test (R = 469, p = .016). Consequently, since motor control and music aptitude were positively correlated, musical sensitivity and auditory abilities accounted for more of the variance in motor coordination than age.

Table 1. Significant correlations of two independent (split half) samples

	melodic perform.	Rhythm perform.	rhythm PMMA	tonal PMMA	total PMMA	K-ABC
Mann-Whitney-U	86.500	93.000	27.500	38.500	29.000	26.500
Wilcoxon-W	191.500	198.000	132.500	129.500	134.000	131.500
Asympt. Significance (2-tailed)	.590	.810	.001	.011	.002	.001

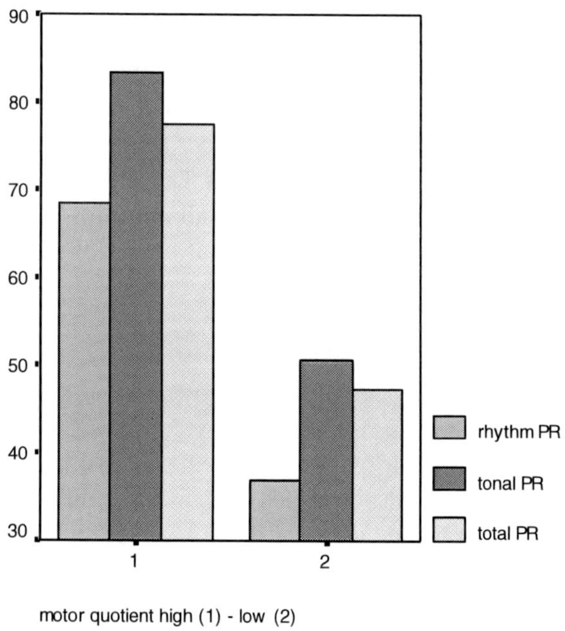

Figure 1. Percentile ranks (PR) of the PMMA-test and subtests (tonal and rhythm) according to the split-half groups of high and low motor quotients.

Table 2. Correlations for MOT motor variables and PMMA music aptitude scores

MOT scores	PMMA scores		
	PR rhythm	PR tonal	PR total
coordination	r =.608** p = .001	.495** .009	.581** .001
fine motor abilities	r = .714** p = .000	.616** .001	.726** .000
balance	r = .635** p = .000	.615** . 001	.636** .000
motor control	r = .527** p = .004	.510** .007	.628** .000

The musical activities in children's leisure time did not contribute to this association, that is, children with and without extra-musical activities were both equally distributed over the total sample. However, the advancement of motor coordination is reflected by the distribution of those children who are engaged in extra-musical activities in their leisure time (singing in a children's choir or learning a musical instrument).

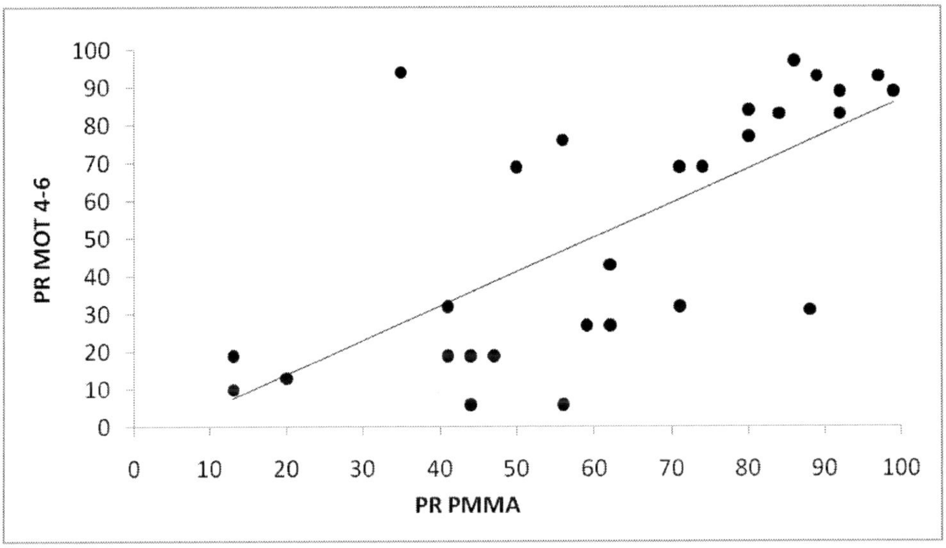

Figure 2. Correlation between the percentile ranks (PR) for movement (MOT 4-6) and music (PMMA). A Pearson correlation of 0.579** is highly significant.

Seventy-nine percent of the children with extra-musical activities were found in the high motor quotient group, whereas only 21% were found in the low motor quotient group. In contrast, those children who more preferentially engaged in motor activities like scramble, or gymnastics in their leisure time were equally distributed amongst the two motor groups (~ 50% in each group). Differences between motor groups for children with musical hobbies (like singing, listening to music, etc.) could not be investigated because all parents reported these hobbies for the children in this study.

The finding that children who engage in extra-musical activities were over-represented in the high motor group is not surprising, since playing a musical instrument requires motor coordination and motor control. However, if we look for differences in the development of motor behavior for children who sing in a choir compared to those who play a musical instrument, the data unexpectedly suggest that singing seems to have a stronger effect on motor development than playing an instrument. However, it should be noted that the number of children who sang in a choir (n = 5) is too small to permit any valid conclusion. A multivariate within-subjects analysis of variance (MANOVA) for the factor "extra musical activities" found a main effect only for the motor quotient ($F = 7.121$. $p = .004$), but not for intelligence or music aptitude.

3. EXPERIMENT II

3.1. Aim of the Study

Because the results of the first behavioral study revealed a correlation between motor control and musical development, the question arises as to whether this might also be reflected by physiological and biomechanical data. Therefore, a second experiment was conducted with the same children using their behavioral data from the first experiment. The rationale for the possibility of a physiological correspondence between music and motor abilities is that muscle activation, balance, and body oscillation might indicate a transformation from spinal to cortical motor control. The implementation of biomechanical measures aims to investigate a possible correlation between music aptitude and the neural foundation of proprioceptive sensitivity. If a relationship can be found, children with high music aptitude could be differentiated from children with lower music aptitude by their motor behaviour in the experimental tasks. The motor test data of Experiment I were thus correlated with the biomechanical measures of Experiment II. Finally, the cognitive scores of Experiment I were correlated with the same biomechanical measures.

3.2. Participants and Measures

A subset of sixteen children from the first sample was chosen. Eight of the children came from the subset of children who had the highest scores on the PMMA in Experiment I, and the other eight children came from the subset that had the lowest scores on this test. In addition to the behavioral data previously collected for Experiment I (motor, music aptitude, cognitive), the following four physiological and biomechanical data were collected:

1. Measurements of the medio-lateral and posterior-anterior displacements during a one-leg stand on the two dimensional, free-floating platform Posturomed®. The test-protocol was performed unilaterally while right-handed children stood on their right leg and left-handed children stood on their left leg. The data provide a measure of postural stability; wider deflections indicate a lower ability to stabilize balance while smaller deflections refer to a more stable balance.
2. Electromyography (EMG) was recorded during a one-leg stand on the Posturomed® from musculus peroneus longus, musuculus tibialis anterior, and musculus soleus in order to measure neuromuscular activity.
3. Two parameters of jumping ability (Squat-Jump, SJ, and Counter Movement Jump, CMJ) were measured on a force plate. CMJ is initiated from a standing position. The child performs a preparatory dip movement and jumps upwards. The reactive strength in a long Stretch-Shortening-Cycle is tested. SJ starts from a static semi-squatting position (knee angle 90°) without any preliminary movement, measuring the maximal concentric jump-height.
4. Body Oscillations, by having the children continuously bend and stretch their knees, were recorded from a Leonardo® Mechanograph. This measure measures the ability to control and coordinate body movements in a rhythmically stable manner.

3.3. Procedure

The children were asked to perform the above tasks in a playful context (staying on one leg like a stork; jumping like a kangaroo etc.). Each child performed three trials that were then averaged. An electromyography (EMG) was recorded during the balance task to measure muscle activity. Activation of the musculus peoneus longus, musculus soleus and musculus tibialis anterior were recorded from the supporting leg by two adhesive bipolar electrodes (Ambu Blue Sensor P, type P-00-S/50, Ag-AgCl) which were fixed at a distance of about 2 cm on the muscle belly of each of the three muscles. A reference electrode was placed on the patella. The raw data were then subjected to an offset-correction and a bandpass filter (Butterworth 4th order bandpass filter 10-500 Hz), and the signals were rectified for further computation of an integral value of muscle activity. These data were used to generate three more parameters indicating the proprioceptive sensitivity of the children. The muscle tension that is needed to stabilize balance is mainly evoked by feedback signals from the proprioreceptors. A higher muscle activity for the same postural displacement indicates better proprioceptive sensitivity. Based on this, a proprioceptive amplification ratio (PAR) was calculated for each muscle. The integrated EMG (iEMG) of the musculus peroneus longus was referred to the medio-lateral direction:

$$PAR_{m} - \frac{iEMG\, M.peroneus\ longus}{\textit{displacement medio-lateral}}$$

For the anterior-posterior direction, the activity of musculus soleus was used in isolation as well as in combination with the musculus tibialis anterior:

$$PAR_{ap_1} = \frac{iEMG\, M.soleus}{\textit{displacement anterior-posterior}}$$

$$PAR_{ap_2} = \frac{iEMG\, M.soleus + iEMG\, M.tibialis\ anterior}{\textit{displacement anterior-posterior}}$$

The resulting PAR quotient serves as a measure for proprioceptive sensitivity; the higher the quotient, the better the proprioceptive sensitivity.

3.4. Results

The upper and lower PMMA achievement groups are clearly, but not significantly differentiated by their biomechanical data. The proprioceptive amplification ratios (PAR) are also obviously different for the two PMMA groups, but they too did not differ significantly because of the large amount of variability in the data (Figure 3).

Similarly, both groups are clearly differentiated regarding their biomechanical measures of the motor tasks. However, both Jump and Body Oscillation exhibited highly significant differences ($p = .003$ and $p = .005$ respectively). Aside from this, PMMA raw scores and PAR quotients exhibited a slight linear progression and confirm a clear association of both auditory and motor measures (Figure 4).

Significant correlations emerged for the music aptitude test (PMMA) and two motor tasks, Body Oscillation (r = -.590, p = .01) and Jump (r = .613, p = .01) for the total sample. A one-way within-subjects ANOVA confirmed that the scores of the PMMA exhibited a significant effect on both Body Oscillation and Jump.

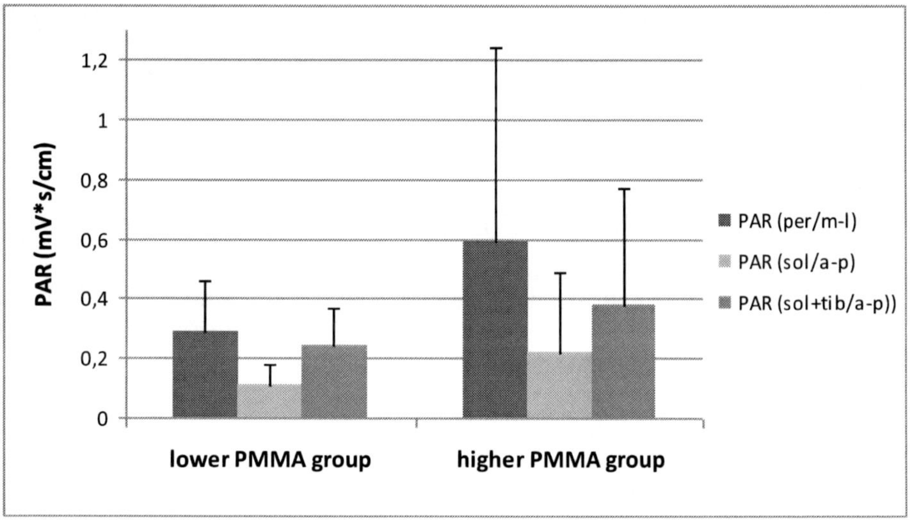

Figure 3. The scores of the two PMMA split half groups yield a clear, but not significant separation with regard to their three proprioceptive amplification ratios (PAR quotients).

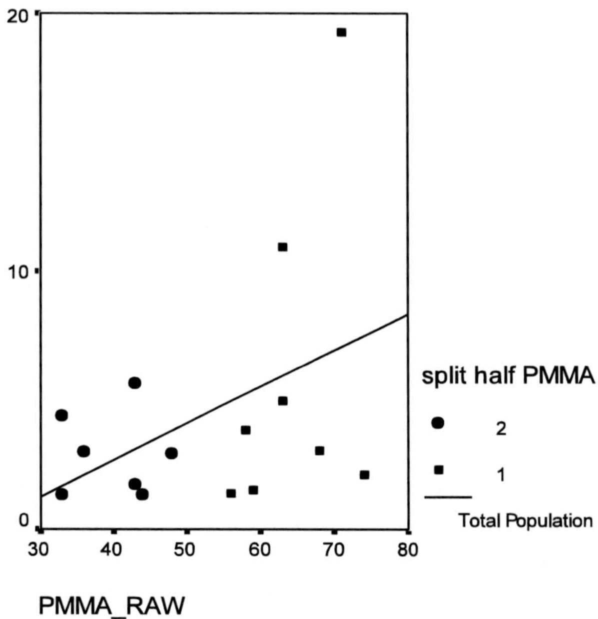

Figure 4. The correlation of musical aptitude (PMMA) and the bio-mechanic amplification ratio (PAR) exhibits a slight linear progression.

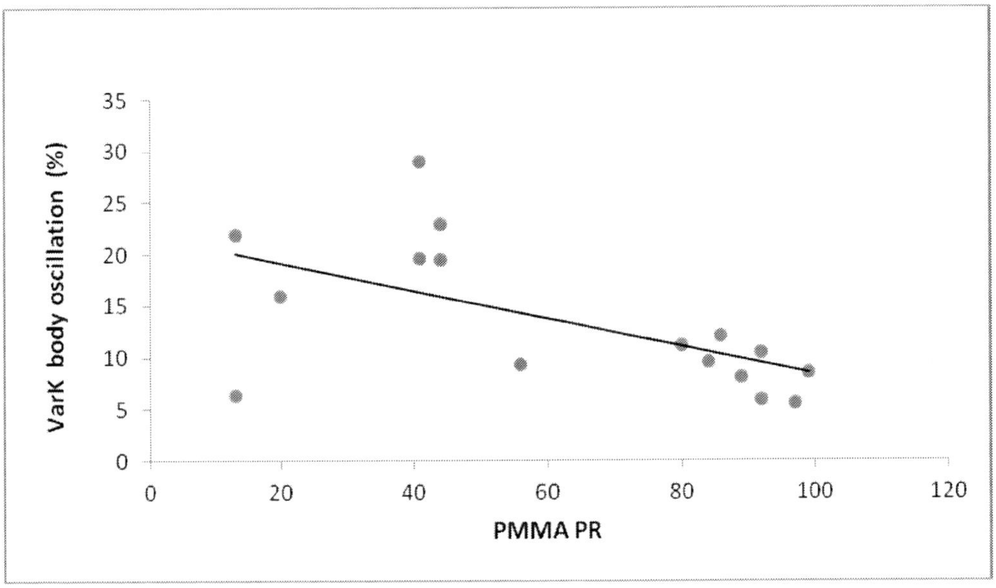

Figure 5. A significant negative correlation (r = – 0.590) is performed for the percentile ranks of PMMA and body oscillation.

Furthermore, the variation coefficients for Body Oscillation exhibited a significant negative correlation with the PMMA values, because small variation coefficients during rhythmic Body Oscillation indicate a more stable and well-balanced variance of the period length. Therefore, a smaller variation coefficient indicates better proprioceptive sensibility. **Thus, the higher the participant's PMMA score, the smaller the variation coefficients** (r = -0.590, $p < .05$, two-tailed; Figure 5).

Since it is reasonable to expect that Jump and Body Oscillation advance with age, the variable age was factored out for all correlations. The results still demonstrate the same effects on a lower level. A clear association exists, but the significance disappears in a partial correlation without age. Finally, the scores of the behavioral motor test (MOT 4-6) were significantly correlated with Jump ($r = .533$, $p = .049$), whereas no significant correlation was found for the other biomechanical measurements or cognitive development (K-ABC).

CONCLUSION

The data from both experiments clearly demonstrate a correlation between motor and musical development. Although this result was not expected, it is not surprising given that several studies suggest that an auditory-motor network is activated in speech and music (D'Ausilio, Altenmueller, Olivetti Belardinelli, and Lotze, 2006; Hickock, Buchsbaum, Humphries, and Muftuler, 2003; Rochet-Capellan and Ostry, 2011) In particular, the inferior colliculus in the midbrain, which receives input from the auditory pathway and the auditory cortex, comprises bimodal neurons that are sensitive to both auditory and sensorimotor stimuli.

Age, of course, also plays an important role in developmental processes. However, once age was controlled for, a significant association between auditory and motor development

Gordon, E. E. (1979). *Primary Measures of Music Audiation (PMMA)*. Chicago: GIA Publ. Inc.

Gordon, E. E. (1997). *A Music Learning Theory for Newborn and Young Children (1990)*. Chicago: GIA Publ. Inc.

Gordon, E. E. (2001). *Preparatory audiation, audiation, and music learning theory. A handbook of a comprehensive music learning sequence*. Chicago: G.I.A. Publ. Inc.

Gordon, E. E. (2007). *Learning sequences in music. A contemporary music learning theory.* (7. ed.). Chicago: GIA Publ. Inc.

Gruhn, W. (1999). The development of mental representations in early childhood: a longitudinal study on music learning. In S. W. Yi (Ed.), *Music, mind, and science* (pp. 434 - 453). Seoul: Seoul Nat. University.

Gruhn, W. (2002). Phases and stages in early music learning. A longitudinal study on the development of young children's musical potential. *Music Education Research, 4*(1), 51 -71.

Gruhn, W. (2010). Body, voice and breath: the corporeal means of music learning. *The Orff Echo, Spring issue*, 34 - 38.

Hickock, G., Buchsbaum, B., Humphries, C., and Muftuler, T. (2003). Auditory-motor interaction revealed by fMRI: speech, music, and working memory in areas spt. *Journal of Cognitive Neuroscience, 15*(5), 673-682.

Hodges, D. A. (2009). Bodily responses to music. In S. Hallam, Cross, I. and Thaut, M. (Ed.), *The Oxford Handbook of Music Psychology* (pp. 121 - 130). Oxford: Oxford University Press.

Jaques-Dalcroze, E. (1977). *Rhythmus, Musik und Erziehung*. (Repr. Basel 1921 ed.). Göttingen: Kallmeyer.

Jarvis, E. (2004). Learned birdsong and the neurobiology of language. In H. P. Zeigler and P. Marler (Eds.), *Behavioral Neurobiology of Birdsong* (Vol. 1016, pp. 749 - 777). New York: The New York Academy od Sciences.

Kaufman, N., and Kaufman, A. S. (2007). *Kaufman assessment battery for children (K-ABC). Deutschsprachige Fassung*. Amsterdam: Swets and Zeitlinger.

Kubo, M., and Ulrich, B. D. (2006). Early stage of walking: development of control in mediolateral and anteroposterior directions. *Journal of Motor Behaviour, 38*(3), 229-237.

Laban, R. v. (1988). *Die Kunst der Bewegung*. Wilhelmshaven: Noetzel.

Malinowski, A. (2000). *Bewegung und Bewegungslernen in Sport und Musik* (Diplomarbeit Universität Tübingen ed.). Steinach: Eigenverlag.

Marler, P. (2000). Origins of music and speech: Insights from animals. In N. Wallin, B. Merker and S. Brown (Eds.), *The origins of music* (pp. 31 - 48). Cambridge MA: MIT Press.

Merker, B. (2005). *Between perception and performance: vocal learning as key constraint on the path to music and language*. Paper presented at the The Neurosciences and Music II, Leipzig.

Mooney, R. (2004). Synaptic mechanisms for auditory-vocal integration and the correction of vocal errors (Vol. 1016, pp. 476 - 494). New York: Annals of the New York Academy of Sciences.

Patel, A. D., Iversen, J. R., Bregman, M. R., and Schulz, I. (2009). Experimental evidence for sychronization to a musical beat in a nonhuman animal. *Current Biology, 19*, 827-830.

Phillips-Silver, J., and Trainor, L. J. (2007). Hearing what the body feels: auditory encoding of rhythmic movement. *Cognition, 105*(3), 533 - 546.

Phillips-Silver, J., and Trainor, L. J. (2008). Vestibular influence on auditory metrical interpretation. *Brain and Cognition, 67*(1), 94 - 102.

Rochet-Capellan, A., and Ostry, D. J. (2011). Simultaneous acquisition of multiple auditory-motor transformations in speech. *Journal of Neuroscience, 31*(7), 2657-2662.

Schachner, A., Brady, T. F., Pepperberg, I. M., and Hauser, M. D. (2009). Spontaneous motor entrainment to music in multiple vocal mimicking species. *Current Biology, 19*(10), 831 - 836.

Sedlmeier, P., Weigelt, O., and Walther, E. (2011). Music is in the muscle: how embodied cognition may influence music preferences. *Music Perception, 28*(3), 297 - 305.

Shusterman, R. (2008). *Body consciousness: a philosophy of mindfulness and somaesthetics.* Cambridge: Cambridge University Press.

Thaut , M. H. (2003). Neural basis of rhythmic timing networks in the human brain. *Annals of the New York Academy of Sciences, 999*, 364 - 373.

Thelen, E., and Cooke, D. W. (1987). Relationship between newborn stepping and later walking: a new interpretation. *Developmental Medicine and Child Neurology, 29*(3), 380-393.

Trevarthen, C. (1999). Musicality and the Intrinsic Motiv Pulse: evidence from human psychobiology and infant communication. *Musicae Scientiae. Special issue, 1999-2000,* 157 - 213.

Winkler, I., Háden, G. P., Ladinig, O., Sziller, I., and Honing, H. (2009). Newborn infants detect the beat in music. *Proceedings of the National Academy of Sciences of the USA, 106*(7), 2468-2471.

Zeigler, H. P., and Marler, P. (Eds.). (2004). *Behavioral neurobiology of birdsong* (Vol. 1016). New York: Annals of the New York Academy of Sciences.

Zentner, M. R., and Eerola, T. (2010). Rhythmic engagement with music in infancy. *Proceedings of the National Academy of Sciences in the USA, 107*(13), 5768-5773.

Zimmer, R., and Volkamer, M. (1984). *Motoriktest für vier- bis sechsjährige Kinder, MOT 4 - 6.* Weinheim: Beltz.

In: Applications in Arts Biomechanics
Editors: Gongbing Shan and Peter Visentin
ISBN: 978-1-62808-015-5

Chapter 3

QUANTIFYING COMPENSATION STRATEGIES BETWEEN GROSS AND FINE MOTOR CONTROL IN VIOLIN PERFORMANCE

Gongbing Shan[1], Peter Visentin[2], Guillaume Tardif[3], Airdrie Robinson[2], Jared Evans[1] and Sarah Crowe[1]

[1]Department of Kinesiology, Faculty of Arts and Science
[2]Department of Music, Faculty of Fine Arts
University of Lethbridge, Canada
[3]Department of Music, Faculty of Arts
University of Alberta, Canada

ABSTRACT

In order for scientific research in music performance to have meaningful application, it must first understand the needs of the vocation and then prove its utility through methods that provide results that are user-friendly. For many musical activities, interaction with an instrument requires significant amounts of motor control training. To date there are few quantitative studies in the area of the movement analysis and biomechanics of music performance.

The goal of this study is to develop a methodology for the purpose of revealing relationships between gross and fine motor control by examining: 1) gross and fine motor control simultaneously, 2) their interaction, and 3) interplay that may be dependent on the tools of performance. A twelve camera, 3-D motion capture system (VICON MX40, 200 frames/s) was used to measure full-body movement of professional violinists using 68 reflective markers on the body, the left hand and fingers, the violin and the bow. Results showed: 1) control pattern changes within single joints, accounting for playing conditions and/or the tools of performance, 2) interactions between joints, including gross to gross, fine to fine, and gross to fine control pairs, and 3) control pattern consistencies of some joints that may be linked to the effects of training.

These suggest that successful performance on the violin requires flexible motor control strategies that may be understood as compensatory in nature. Understanding of the role of compensation as evocable strategy rather than a default condition has great potential to improve music pedagogy and vocational outcomes for performers.

INTRODUCTION

Instrumental music performance is acknowledged to be an activity requiring complex brain, behavior and motor control development. In the 21st-century, there has been increasing interest in understanding the responses of the brain to musical stimuli (Doidge, 2007; Huron, 2006; Levetin, 2007; Snyder, 2000). For a performer, not only is the brain responsible for responding to external musical stimuli, it is also charged with the tasks of creating them in a **manner that "responds to the moment"** of the performance. Such abilities take significant amounts and training. For many musical activities, bodily movements controlling interaction with an instrument are at the center of music pedagogy. Considering the seeming importance of motor training in this regard, it is surprising how few studies exist that examine gross or fine motor control during performance. Even more surprisingly, a search of the literature reveals no quantitative studies examining the interaction between gross and fine motor coordination in music performance. Hence the goal of this study is to develop and document a methodology for the purpose of revealing relationships between gross and fine motor control in music performance.

In most physical activities, motor control patterns exhibit the characteristic wherein either gross or fine motor control appears to be dominant. In much of daily living, and in most athletic training for sports, it is reasonable to conceive that the majority of activities (e.g. walking, running and lifting) mainly rely upon large muscles (gross motor control) where smaller muscles function in significant stabilizing roles. Fewer activities, like writing, typing, or using tools and devices such as chop sticks or the iPhone, rely mainly on smaller muscle group coordination (fine motor control) where gross motor control supplies foundational support or is nearly arrested. In the case of much instrumental music performance (e.g. playing the piano or the violin) there is visibly co-dependent interaction of both gross and fine motor control with dynamic interchange of roles depending on the musical context. To complicate matters further, because music unfolds over time, during which all motor control must be guided, performers' **motor behaviors** must be part of a contextual, forward-planned process. At the same time, since it is undesirable for performers to merely act as mechanical executors of instructions on a printed page (von Hasselbach, 2012), they must maintain sufficient flexibility to adapt to and accommodate circumstantial developments. At the highest levels of competency, adaptation will include immediate forward *re*-planning. Thus, music performance must be considered a process, not a reconstruction; a reality that creates significant, but not insurmountable, challenges for researchers who apply scientific techniques to fine arts performance.

Another confounding factor for movement analysis in arts performance is that the **physical dimensions of a performer's body (anthropometry) influences** motor control. In great part, the personal **'style' of a performer** derives from his or her development of a variety of motor behaviors that are compensatory in nature; accommodating body size and shape as well as producing outcomes that he/she would perceive as falling within an acceptable range of musico-cultural/aesthetic **"expectations"**. This, in turn, poses particular challenges for both research and pedagogy. An astonishing example of motor compensation can be observed in the performance of Canadian violinist Adrian Anantawan (Wikipedia, 2012). Adrian was born without a right hand, no wrist and only a partial forearm. This means he has only two arm segments with two functional joints in his bowing arm – the arm responsible for much, if not most of the violinistic nuances associated with articulation and expression. A "normal" violinist would have the use of three large segments in the arm (upper arm, forearm, hand) plus the fingers (with a total of 17 functional joints) for the finest motor adjustments. Astoundingly, Adrian has found compensatory motor control mechanisms that allow him to accomplish musical ends that might be expected at the highest of professional levels (Anantawan, 2012). His accomplishments highlight that there are many motor control means to the same end. For meaningful motor control research in music performance and pedagogy, research methodology must be developed to examine its many facets simultaneously and in context.

Literature Review

Research in this regard first needs to identify the kinds of tools that can examine details of both gross and fine motor control. To date there are few quantitative studies in the area of the biomechanics of music performance. The most recent of these have adopted research tools from the field of kinesiology and movement science, adapting them to musical contexts. The vast majority of such studies focus on quantification of gross motor control mechanisms; far fewer examine fine motor behaviors (Ferrario, Macri, Biffi, Pollice, and Sforza, 2007; Iltis and Givens, 2005; Palmer, 2001; Schoonderwaldt and Altenmüller, 2011; G. Shan and Visentin, 2003; Turner-Stokes and Reid, 1999; Visentin and Shan, 2003). Other research examines motor control indirectly. Quantification of the movement of equipment used in an activity may be linked to the motor control of the performer (Schoonderwaldt and Altenmüller, 2011). No quantitative studies examine both gross and fine motor control simultaneously. By showing relevant results, each of the exemplary studies cited above provides proof of both the viability of adopting the tools of science in a music performance context and of its methodological utility. Just like sports research conducted during the 1980s, the first steps of music performance research have been to examine basic skills as a foundation for questions of greater complexity.

For music performance, 3-D motion capture has become the preferred tool for biomechanics research (Ferrario, et al., 2007; Visentin and Shan, 2003). 3-D motion capture uses multiple cameras to register anatomic landmarks such as joints. To do so, researchers place reflective markers on subjects' bodies in identified locations and the cameras triangulate and record movement data in XYZ coordinate computer space. From these coordinates, researchers can track movements of the markers, and more importantly create models of subjects' musculoskeletal characteristics. Since computer models are vital for quantitative analysis of captured data, the quality of a model is a fundamental determinant of the quality of research.

To date, nearly all existing biomechanical models used in music research have been adapted from three main sources: 1) sports and movement science, 2) ergonomics, and 3) anatomic models designed in the abstract. Studies using each of these sources highlight their utility and provide foundation for developing music-specific models. It is important to note that all models necessarily simplify reality, so a key determinant of a model's utility lies in clear definitions of the goals of research. In order to analyze complicated movements, target features must first be identified. Although simplification is required to make models practical to use, it must not go so far as to permit pre-filtering of results based on researchers' presuppositions about what is important. In music research, models analysing fine motor control have been a stumbling block. As such, existing full-hand biomechanical models from the sources above provide a valuable starting point for integrated research that examines both gross and fine motor control. The examples below may be considered representative of hand models found in our literature review and discussion of each will demonstrate the difficulties of adapting existing models to music research.

Ergonomics Model

At first glance, ergonomics models that explore mechanical functions of the hands and fingers appear to be likely candidates for adaptation to a music research context. They are designed for analysing a variety of activities such as computer keyboarding, reaching, and grasping. In 2005, Cerveri and his colleagues used motion capture technology with 40 surface markers to develop a 3-D biomechanical model for determining kinematic characteristics of hand skeleton (Figure 1) (Cerveri, Lopomo, Pedotti, and Ferrigno, 2005). The model was designed to analyse comprehensive biomechanics of workers who use powered hand tools or engage in manual assembly operations.

Such tasks are linked to occupational injuries such as carpal-tunnel-syndrome and hand–arm-vibration-syndrome (NIOSH, 1997); not incidentally, the high rate of occupational musculoskeletal injuries among performers is a pressing reason for undertaking such research in a music context. **Cerveri's research examined** the hypothesis that injury severity could be related to gripping mechanics. The model design neglected individualized control of the knuckles for very good reason – gripping usually uses bundled, simultaneous action of the knuckles. If the movement of one knuckle is known, the movement of all of them may be inferred. The limited utility of this design in a context where individualized finger control is necessary renders it problematic for music research.

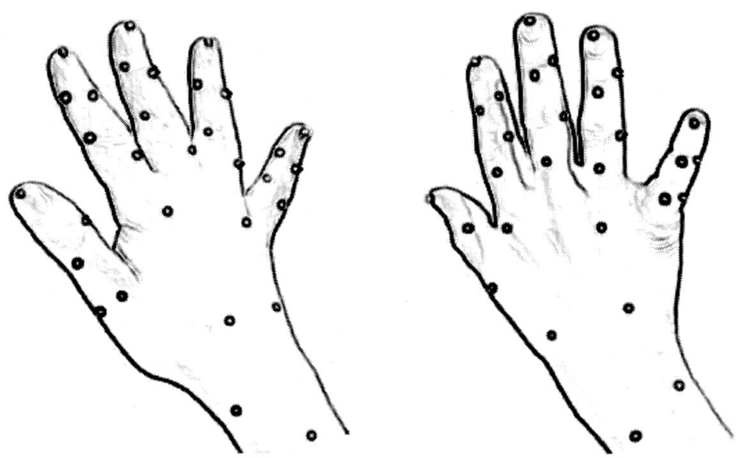

Figure 1. The marker set for gripping mechanics modeling (Cerveri, et al., 2005).

Anatomic Model

One of the most detailed hand models documented in the literature employs almost 90 motion capture markers on a single hand (Figure 2) (Buczek et al., 2011). By using models with this degree of complexity, it becomes possible to examine every detail of hand and finger movement including abduction, adduction, and rotation of each finger segment. Buczek's model included enough complexity to nearly eliminate skin movement across the metacarpals so that it "would not corrupt inverse dynamics calculated for the phalanges" (Buczek, et al., 2011). However, even though it may seem intuitively commonsense, it is not necessarily true that "more complicated" models are always "better". The motion capture marker set uses up to six markers on each finger segment (18 markers on a single finger). This is impractical for use in many activities, music performance included, because markers can interfere with movement; more markers mean more potential for interference. If one imagines violin playing, the left hand fingers curl as they meet the instrument's fingerboard to sound pitches, they operate in very close proximity to one another and frequently rub. Further, the neck of the violin is cupped in the same hand. Inevitably, markers placed on the inside of fingers for such an activity would be entirely occluded (obscured from the view of the cameras) even if they didn't interfere with finger movements. Clearly, a simplification of this Buczek's would be required to render it functional for violin research. The question for researchers is whether or not details lost by simplification will prove meaningful. In a music performance context, it is questionable whether all segments of the fingers need be measured for rotation (which is one of the main purposes of having three or more markers on each segment).

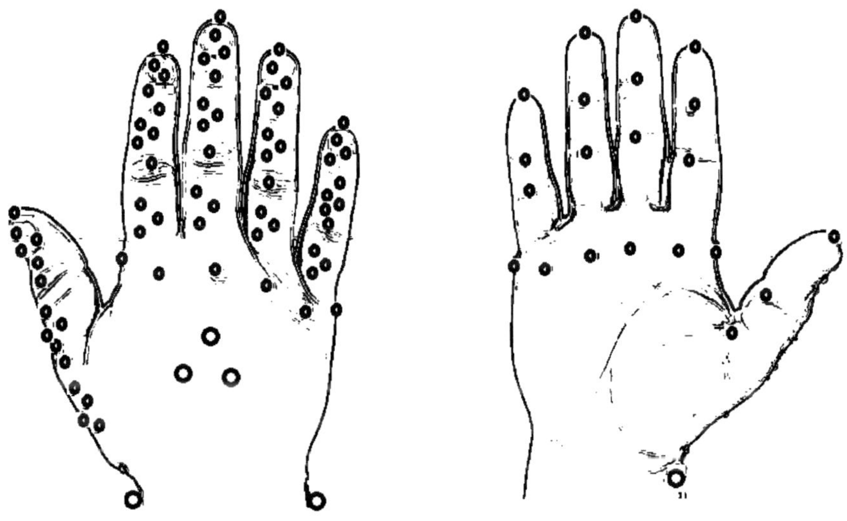

Figure 2. The marker set for inverse dynamic modeling to determine the anatomical functions of fingers in medical applications (Buczek, et al., 2011).

The joints between the 1st and 2nd and 2nd and 3rd phalanges act mainly as hinges, where flexion and extension dominate the movement to such a degree that it becomes reasonable to ignore both abduction and adduction (Ab/add) and rotation (Rot). Clearly, for music performance research, the marker-set of Buczek is overly complicated, would suffer from problems of marker occlusion (cameras not being able to see the markers) during any except open-handed activities and, further, challenges the current limits of motion capture technology to process raw 3-D data.

Movement Science Models

Modeling in movement science is usually based on the needs of the investigators. Depending on the research questions being asked, models may require different degrees of complexity or marker orientation. One model found in the music research literature comes from a piano study where fingertip movement was examined (Ferrario, et al., 2007). By using 7 markers on a single hand (Figure 3) the researchers found that during playing, pianists could use different techniques to reach the same results. Although the model allowed researchers to establish alternative control patterns, details related to the control of the individual finger joints was missing. It would take considerably more detail to begin to answer the investigators' speculative assertion that "the actual effect of extraneous motions on the body should be assessed to verify their relationship to overuse injuries" (Ferrario, et al., 2007).

In another movement science application, Metcalf and colleagues (2008) proposed a marker-placement (26 markers) which allowed the calculation of the composite movements of the wrist, dorsal aspect of the transverse metacarpal arch, fingers and thumb (Figure 4). Their method was not developed as a complete biomechanical model and, unlike the examples above, is not suitable for use in clinical practice. The focus was on the analysis of flexion/extension of individual fingers during lifting (Metcalf, Notley, Chappell, Burridge, and Yule, 2008). Tests of the model ascertained that research results were repeatable and accurate to within one degree in all directions. Many of the merits of this model prove to be important in music research – relative simplicity, marker placement that avoids occlusions and minimally interferes with activity, and repeatability.

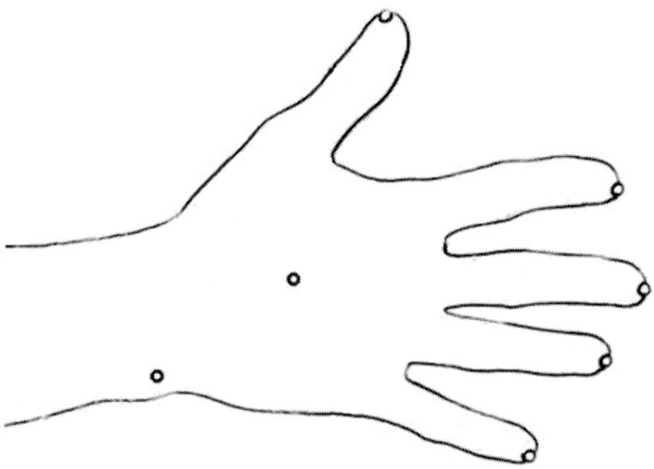

Figure 3. The marker set for a study of finger movement during piano performance (Ferrario, et al., 2007).

Each of the digits has the possibility of three degrees of freedom: flexion/extension, ab/adduction, and rotation. In this research, only the thumb was modeled for rotation. However, we can agree with their motivation. At one point, more complexity does not produce more useful results. Similarly to Cerveri's (2005) model for gripping, Metcalf's model appears to consider rotation of individual fingers to be unimportant for lifting tasks.

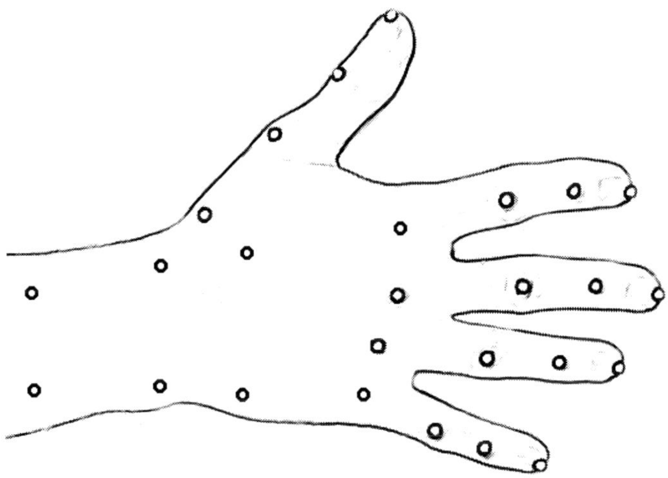

Figure 4. The markers' set-up for quantifying flexion/extension and abduction/adduction of fingers' movements (Metcalf, et al., 2008).

In summary, models from sports and movement science, ergonomics, and anatomic research can each contribute to research of complex activities not specifically intended by the models' designers. It is important to be able to identify the strengths and weaknesses of a number of models in order to begin developing functional models suitable for research in instrumental music performance. In addition to the concerns discussed above, two other factors must be considered. Realistically, music performance is a "full body" art; an isolated examination of the hands will necessarily have limited utility because in a complex kinematic chain (e.g. the arms and hands) there are multiple compensatory means to the same end. Second, models must take into account that, in instrumental performance the performer is interacting with a piece of equipment.

To date, almost all biomechanics research has employed models exclusively examining either gross or fine motor control. A partial reason for this may be technological; it has only been about a decade since motion capture has reached the state where it can, in all practicality, simultaneously record both gross and fine control data. The design below extends the utility of a full-body 15 segment model that examined gross motor control (G. Shan and Westerhoff, 2005). To this basic model, we attached a 16-segment left-hand (fine motor control) model. Since our research goal was to examine the interaction between gross and fine motor control mechanisms during violin performance, we began with the left hand only because finger movement is both visibly apparent and, to great degree, dependent on the physics of producing a sound on the instrument (e.g. pitch and tone control). The method that follows tests a full-body model that considers: 1) gross and fine motor control, 2) their interaction, and 3) interplay that may be dependent on the tools of performance.

METHOD

A 3-D motion-capture system was used to measure full-body movement using 68 reflective markers - 39 on the body, 22 on the left hand, 4 on the violin and 3 on the bow. A twelve-camera VICON MX40 motion capture system (VICON Motion Systems, Oxford Metrics Ltd., Oxford, England) tracked the markers at a rate of 200 frames/s. Figure 5 shows a 3-D computer reconstruction of a single trial including camera placement, capture volume, and a rendered stick figure of a test subject. So far, more than 250 trials of seven subjects have been used to test this model. Protocols for the tests were taken from standard exercises and pedagogical literature. Subjects were all of professional level, having between 18 to 46 years of experience studying and performing on the violin. The figures and discussion below are taken from a single subject since the purpose of this paper is to elaborate methodology. The description of each portion of the model is given separately below.

Body Model

Thirty-nine body markers (9 mm. in diameter) were used to build a 15-segment full-body biomechanical model. Markers were placed on subjects as follows: 1) four on the head, 2) trunk markers on the sternal end of the clavicle, xiphoid process of the sternum, C7 and T10 vertebrae, right scapula, left and right anterior superior iliac, posterior superior iliac, 3) upper-extremity markers on the right and left acromion, lateral side of upper arm, lateral epicondyle, lateral side of forearm, styloid processes of radius and ulna, and distal end of 3rd metacarpal bones, and 4) lower extremities markers on left and right lateral sides of thigh and shank, lateral tibial condyle, lateral malleolus, calcaneus and big toe.

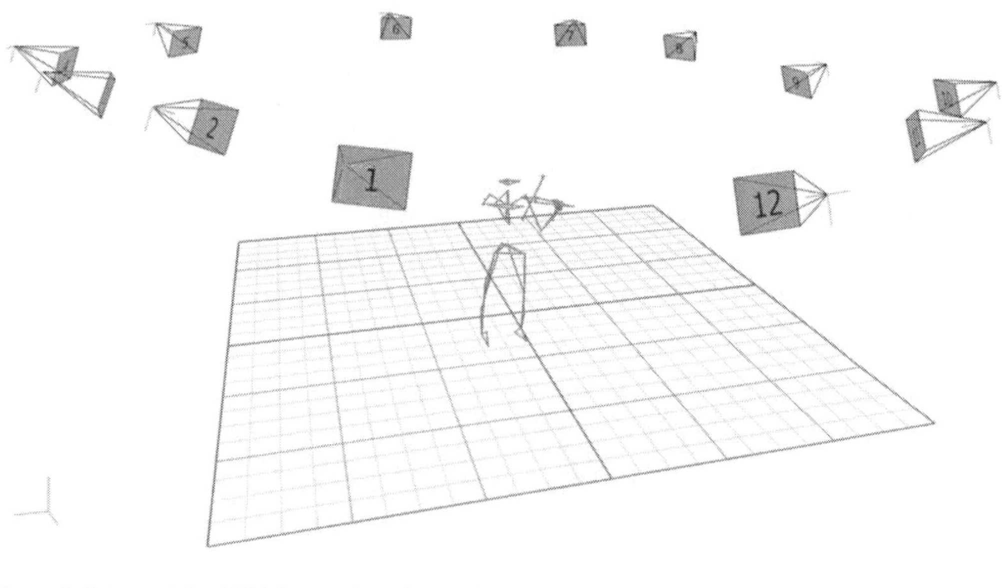

Figure 5. Set-up of the 12 high-speed motion capture cameras.

Calibration residuals **were determined in accordance with VICON's guidelines** and yielded positional data accurate within 0.4 mm. Kinematic data collected using this marker set supplied information such as marker position at all times, changes of position, velocities, and accelerations, as well as joint angles and ranges of motion (ROMs). These were used to quantify gross and fine motor control.

Hand Model

Twenty-two left-hand markers were placed as shown in Figure 6. Spherical markers 4.0 mm in diameter were placed on the hand and finger joints. Finger tips had markers of mixed sizes with 3.0 mm semi-spherical markers on the index and middle fingers (which because of the physics of playing the violin, tended to rub on each other) and 4.0 mm spheres on the rest. 4.0 mm markers were used wherever possible on the hands because 3.0 mm semi-spherical markers increase post-processing requirements.

Although similar in appearance to the marker set of Metcalf and colleagues (2008), the above hand model has several significant differences. First, a reference marker was placed on the ring finger to anticipate future studies that may capture data for left and right hands simultaneously (Figure 6).

An asymmetrical marker set is required for computer-driven discrimination between and among markers. Second, a linking marker was needed toward the distal end of the third metacarpal in order to join the hand to our 15-segment body model (Figure 6). Third, and most importantly, the knuckles were modeled as ball joints with three degrees of freedom (flexion/extension, ab/adduction, and rotation).

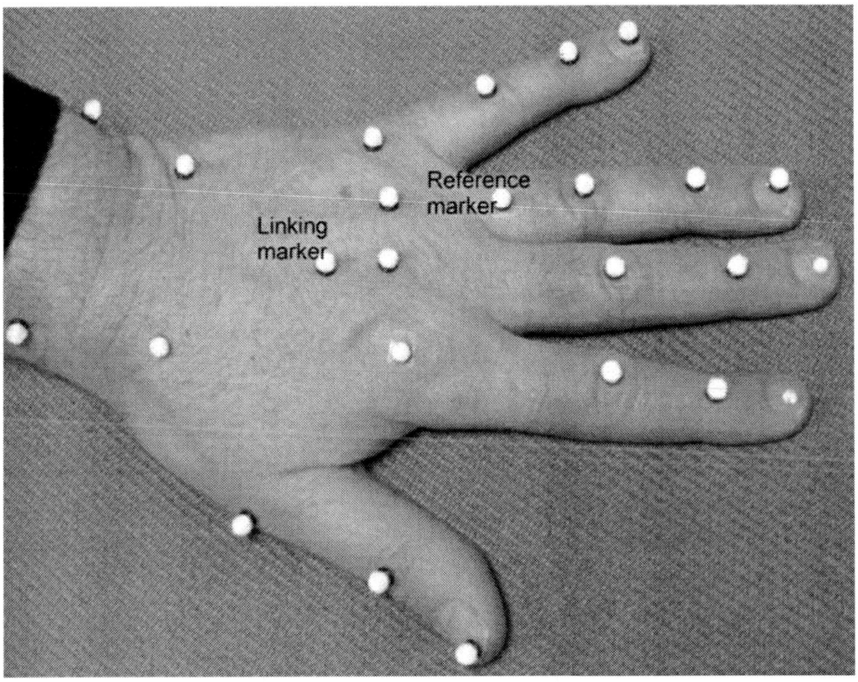

Figure 6. Marker set for the hand in an integrated full body model.

Whereas other models either neglect one or more of these components or, alternatively, require additional markers on the fingers in order to obtain them, our model permitted their calculation from the given marker set. To do so, we made a necessary simplification; we treated the joints below the knuckles as though they were simple hinges. Through calculations using the all markers on the fingers, this model simplification can assign rotational activity to the knuckles.

Violin and Bow

An additional feature of our model design is that it includes data on the interaction of the performer with the tools of performance. The violin and bow are each modeled and linked to the body as separate segments. Markers (4.0 mm) on the violin were placed one each on the tailpiece and scroll, with two on the bridge end of the fingerboard. This orientation established a frame of reference for quantifying the interaction of both the left hand and the bow with the violin. The bow used three markers (one on the button and two on the head) which established its length-wise orientation and its rotation as the violinist plays (Figure 7). Since violin bows are very light (typically around 60 g) and small changes in their physical characteristics can have large effects on playing, the use of small, light markers weighing less than 0.3 grams in total did not influence the balance of the bow or "feel" of playing.

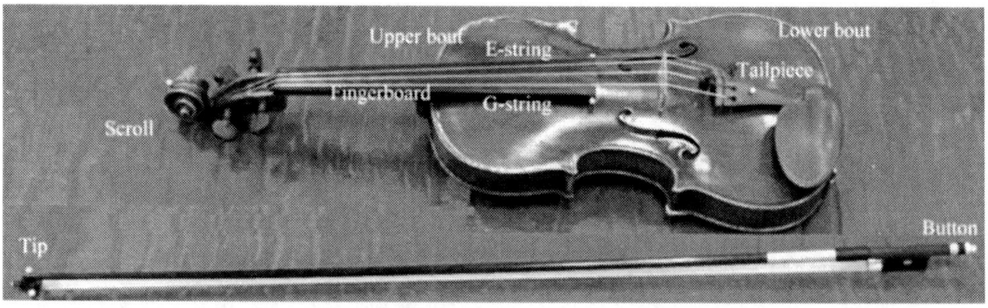

Figure 7. Violin and its parts, as well as marker set.

Data Collection and Processing

A 12-camera setup allowed us to use a relatively large capture volume of 3×3×2.5 meters which permitted considerable freedom of movement for the subjects without negatively influencing accuracy of data.

The motion capture setup described, using small markers as required and larger ones where practical, ensured subjects' movements within the capture volume remained as close to their normal "style" as possible. Raw kinematic data was processed using a five-point (1-3-4-3-1 function) smoothing filter and the resultant data was input into the 15-segment biomechanical body model. Segment lengths, joint angles, and ROMs of shoulders, elbows, hips, knees and ankles were determined. Model segments were identified as follows: head, upper trunk, lower trunk, upper arms, lower arms, hands, thighs, shanks and feet. In such biomechanical modeling, inertial characteristics of the body are estimated using anthropometric norms found through statistical studies (G Shan and Bohn, 2003; Winter, 1990). Figure 8 shows a still frame of one trial and our 3-D computer reconstruction of the skeleton.

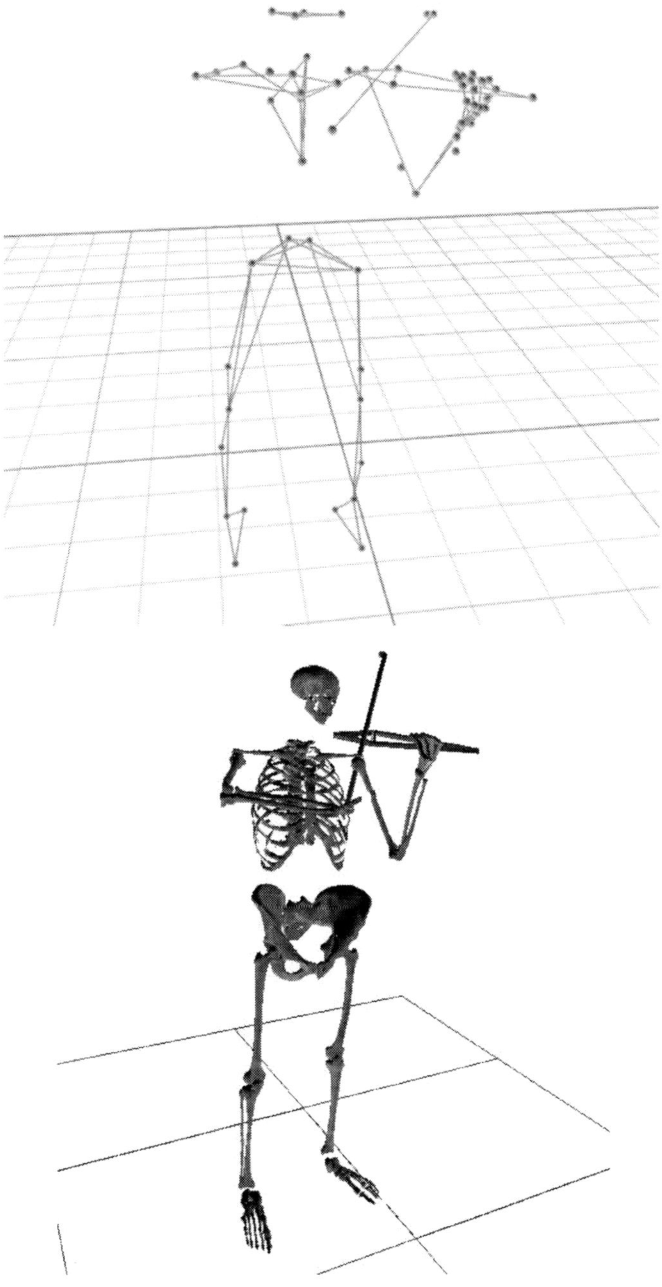

Figure 8. Left: a digitized stick figure using 68 markers: 39 markers for full body model, 22 for left hand, 4 for violin and 3 for bow. Right: A 3-D, 32-segment, biomechanical model derived from the captured markers: 15 segments for body, 19 segments for left hand, one segment each for violin and bow.

RESULTS

Results indicate that it is indeed possible to find relationships, co-dependencies and compensatory behaviors between fine and gross motor control for complex activities using the full body model identified above. The protocol of the exemplary trials below involve fingering the first four notes of a major scale pattern starting on the index finger on the G-string, seamlessly continuing the protocol on all four strings of the violin using all four fingers in sequence on each string, reaching an apogee on the 4th finger on the E-string, then returning to the original G-string pitch using all fingers in reverse sequence (Figure 7). No extended techniques (e.g. shifting during individual trials) were employed. This protocol was used starting on each chromatic pitch on the G-string, from the A-flat below middle C (the lowest sounding fingered pitch on the G-string) to the A-flat above middle C.

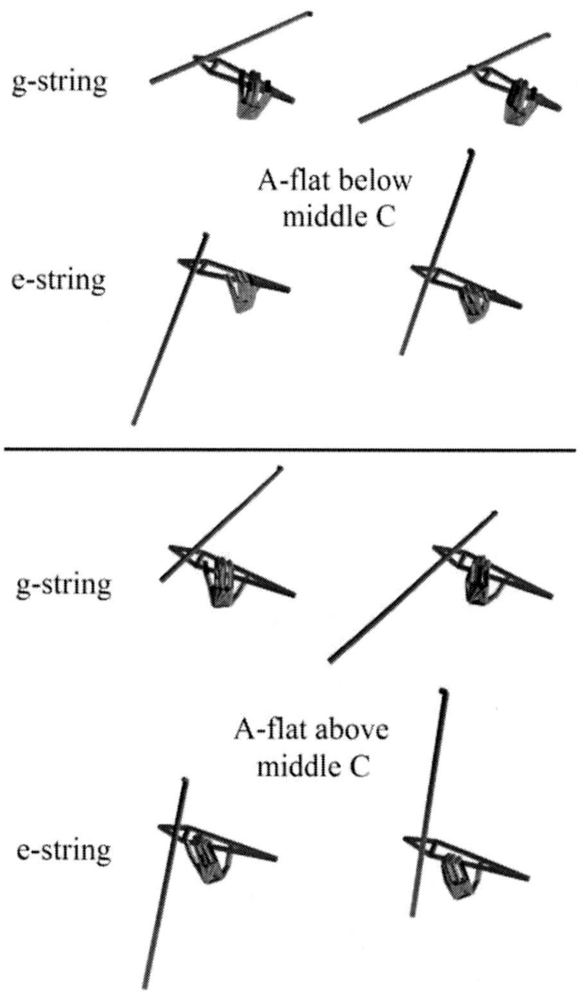

Figure 9. A visual rendering of 1st (left column) and 4th (right column) finger positions on G-string and E-string for two trials: A-flat below middle C (top) and A-flat above middle C (bottom).

Table 1. Average angle of selected joints and their ranges of motion (ROMs) during test trials

			A-flat (below middle C	Middle C	D	A-flat (above middle C
Average Joint Angle (°)	Left Arm	Shoulder (Flex/ext)	23.5	21.8	26.6	27.2
		Shoulder (Ab/add)	-24.4	-23.2	-17.4	-20.2
		Elbow (Flex/ext)	57.2	50.1	49.2	48.3
		Wrist (Flex/ext)	132.9	136.2	146.5	152.4
	Left Thumb	1st Joint (Flex/ext)	121.9	120.6	115.2	111.4
		1st Joint (Ab/add)	-40.1	-39.1	-37	-33.3
		2nd Joint (Flex/ext)	123.1	124.8	130.3	132.8
		3rd Joint (Flex/ext)	156.4	159.2	160.3	163.1
	Left Index Finger	1st Joint (Flex/ext)	172.1	169.5	166.9	162.8
		2nd Joint (Flex/ext)	104.6	94.9	95	96.4
		3rd Joint (Flex/ext)	113.5	123.4	141.3	153.9
	Left Middle Finger	1st Joint (Flex/ext)	153.1	151.3	157.7	156.6
		2nd Joint (Flex/ext)	107.5	104	112	110.4
		3rd Joint (Flex/ext)	114.1	115.7	119.4	117.8
	Left Ring Finger	1st Joint (Flex/ext)	140.2	148.5	152.2	157.4
		2nd Joint (Flex/ext)	114.7	115.5	123.1	124.9
		3rd Joint (Flex/ext)	132.5	134.3	137.8	140.9
	Left 4th Finger	1st Joint (Flex/ext)	113.5	131.3	139.1	146.4
		2nd Joint (Flex/ext)	143.3	142.7	147.7	150.6
		3rd Joint (Flex/ext)	144	145.9	152.5	157.2
ROM (°)	Left Arm	Shoulder (Flex/ext)	20	19.2	21.8	21.4
		Shoulder (Ab/add)	12.3	13.7	16.6	19.3
		Elbow (Flex/ext)	8.9	3.3	4.3	5.1
		Wrist (Flex/ext)	12.5	7.9	7.8	8.2
	Left Thumb	1st Joint (Flex/ext)	8.3	7.9	9.1	7.7
		1st Joint (Ab/add)	13.5	10.5	9.3	6.9
		2nd Joint (Flex/ext)	10.1	7	9.8	11.7
		3rd Joint (Flex/ext)	15.3	10.2	9.9	10.9
	Left Index Finger	1st Joint (Flex/ext)	10.6	15.7	16.5	19.8
		2nd Joint (Flex/ext)	27.5	24.2	26.1	23.7
		3rd Joint (Flex/ext)	14.6	25.3	32.3	38.4
	Left Middel Finger	1st Joint (Flex/ext)	24.6	25.6	23	23.3
		2nd Joint (Flex/ext)	37.3	21.7	33.9	29.5
		3rd Joint (Flex/ext)	35.6	30.6	36.1	34.3
	Left Ring Finger	1st Joint (Flex/ext)	45.1	37.4	30.4	32.2
		2nd Joint (Flex/ext)	25.9	23.8	21.5	26
		3rd Joint (Flex/ext)	24	34.6	24.8	32.1
	Left 4th Finger	1st Joint (Flex/ext)	36.7	24.4	33	24.5
		2nd Joint (Flex/ext)	41.9	38.5	47.8	33.5
		3rd Joint (Flex/ext)	66.2	76.2	67.6	62.4

Figure 9 is a rendering of the left hand, fingers, violin fingerboard and bow using data from 3-D motion capture. Fine motor control changes, clearly visible in the figure, are documented in Table 1 and will be elaborated in discussion thereof. Table 1 lists average joint angles and ranges of motion for four starting pitches: A-flat, middle C, D, and A-flat (one octave above the first A-flat).

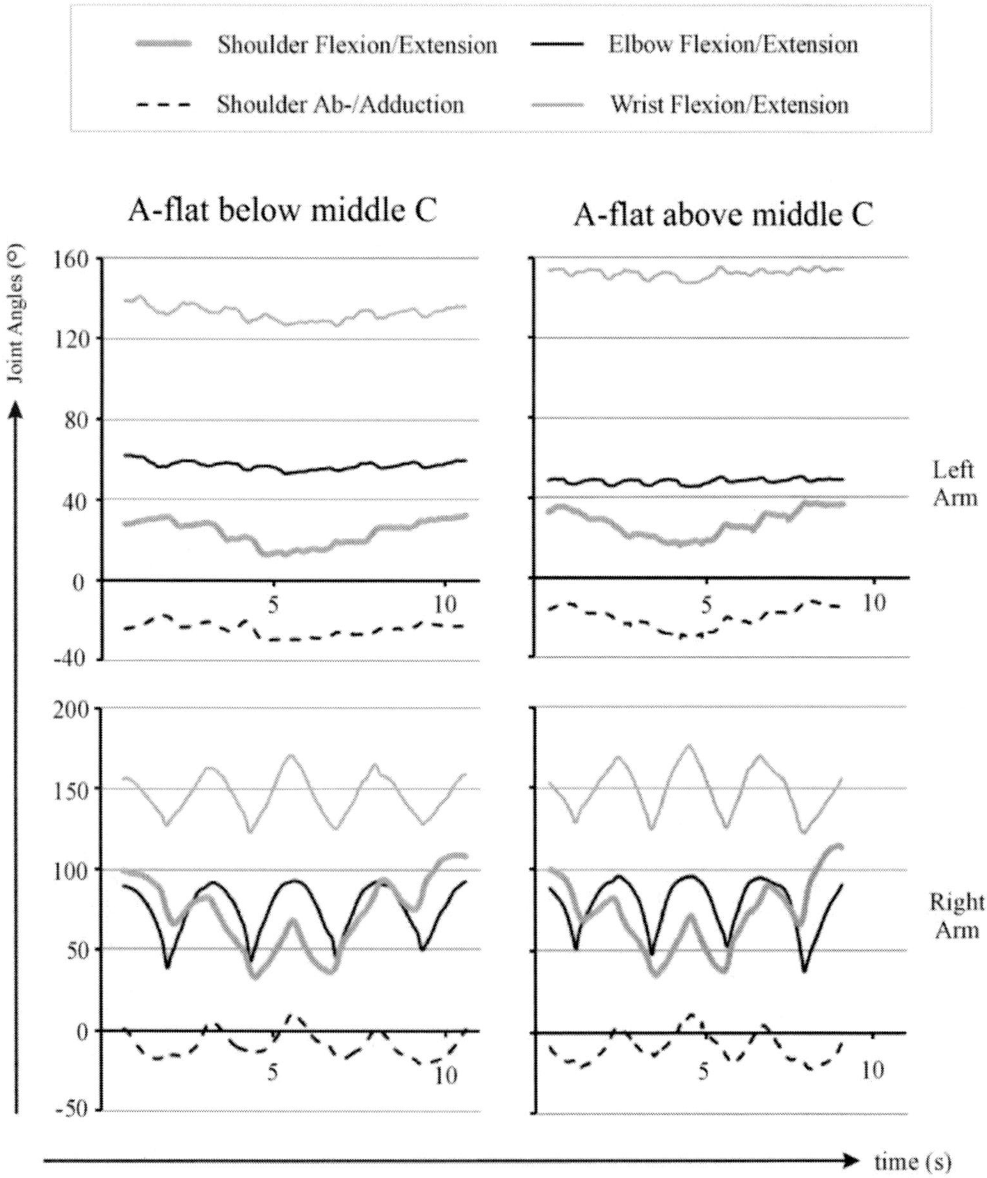

Figure 10. Gross motor as evidenced by joint control of both arms. The trial protocol begins and ends on the G-string, with E-string in the middle of the trial (string order: g-d-a-e-a-d-g). Left and right columns show data for trials beginning on A-flat below middle C and A-flat above middle C respectively.

Finger and thumb joints are identified proximal to distal (starting with the knuckles) as joints 1, 2, and 3 respectively. Average joint angles within each trial are given in the top portion of the table. Ranges of motion (ROM) for each joint (maximum minus minimum joint angles in a single trial) are documented in the lower part of the table.

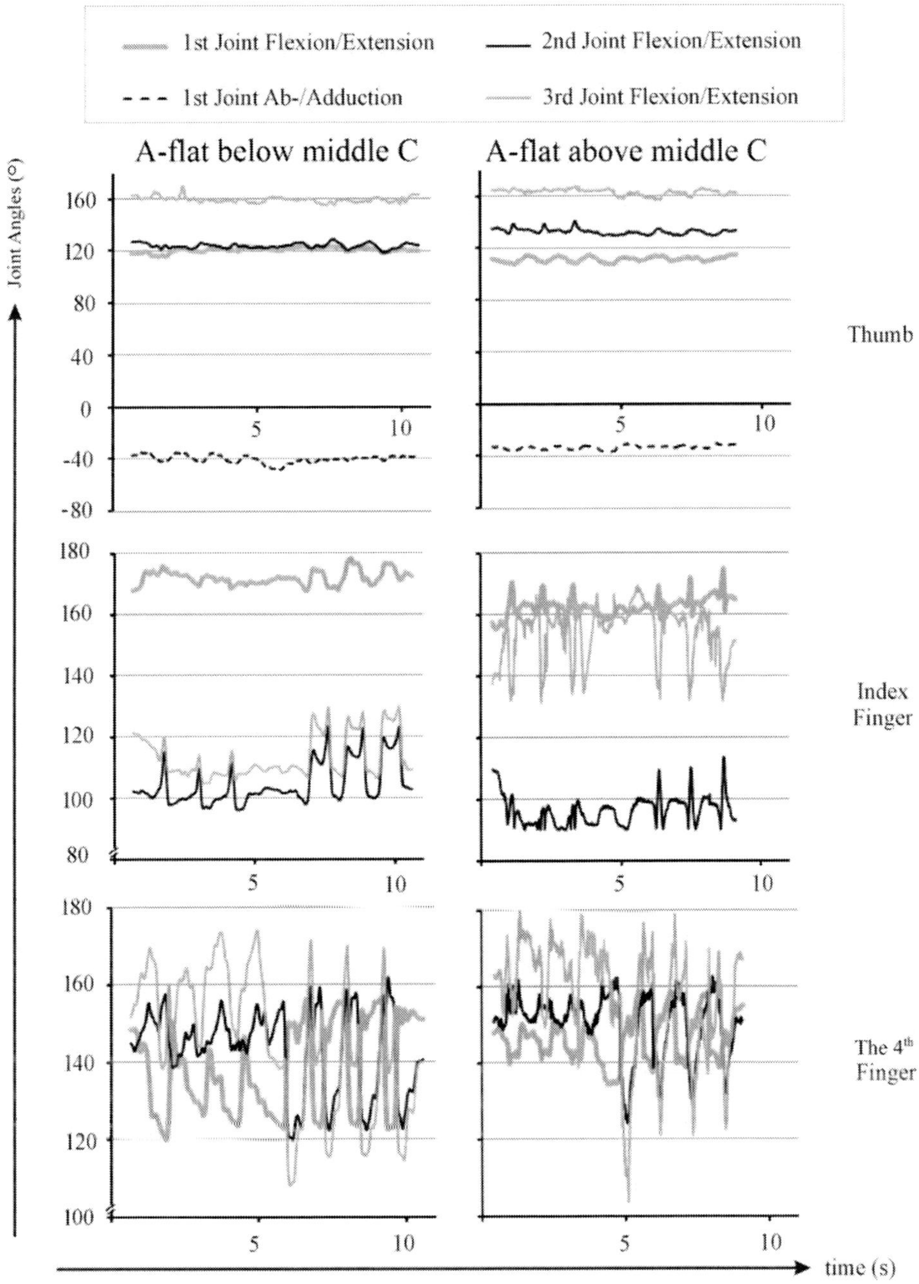

Figure 11. Fine motor control of all joints – thumb, index and 4th finger of the left hand (top to bottom rows, respectively). Left and right columns show data for trials beginning on A-flat below middle C and A-flat above middle C respectively.

For the purposes of introducing the method to the reader, our analysis below uses absolute differences as the measure of significance because relative percentage changes may overstate significance for numbers with small value; for example, in the second row of data (shoulder Ab/add) the percentage change between columns 1 and 4 is over 20%, but the absolute change is only 4.2°, a measure which is too small to have any pedagogic or practical utility. Lines highlighted in dark grey indicate absolute differences of 20° or more to exist between columns 1 and 4 of the data (the extremes of our trials) for the variable identified in the row (e.g. Wrist Flex/ext). Light grey rows indicate differences between 10° and 20°.

For motor control related to overall position and posture (Average Joint Angle), highly significant variances (highlighted in dark grey) can be observed in flexion/extension of: wrist; index finger– 3rd joint; 4th finger – 1st joint. Significant differences (highlighted in light grey) can be seen in flexion/extension of: thumb – 1st joint; index finger – 1st joint; ring finger – 1st and 2nd joints; and 4th finger – 3rd joint. For motor control related to the dynamism of playing, (joint ROMs) highly significant variances (dark grey) occur in flexion/extension of the Index Finger (3rd joint). Significant changes (light grey) occur for flexion/extention of 1st joints of index, ring and 4th fingers (Table 1).

In Figure 10, gross motor control of both arms is graphed for the two conditions of the protocol that are furthest apart in terms of pitch (A-flat below and A-flat above middle C). The playing protocol begins and ends on the G-string, with E-string in the middle of the trial (string order: g-d-a-e-a-d-g). Reading across rows of the figure, a comparison of the various components of joint control can reveal postural compensations due to differing conditions of the trials. Notably, elbow (Flex/ext) and wrist (Flex/ext) show significant changes of joint angle over each trial that clearly can be attributed to postural compensation. Further, the shoulder (Flex/ext and Ab/add) shows subtle changes which appear to be related to the order in which the strings of the violin are played. Interestingly, right arm control (the bow arm) charted in the bottom pair of graphs remain nearly identical.

Figure 11 examines select components of fine motor control in the left hand. Thumb (top row of graphs) shows postural compensation mainly evident in the 1st and 2nd joints. Both index and 4th fingers show compensatory behaviors all of the joints.

DISCUSSION

Until now, the absence of integrated biomechanical models has limited the potential of 3-D motion capture for all contexts where gross and fine motor control have functional yet variable codependency – a condition that is clearly the case for instrumental music performance. Motion capture, first developed for sports and clinical settings, has a long history of successfully examining either gross or fine motor control, but not both. In sports, successful outcomes are typically defined in terms of quantifiable measures (e.g. run faster, or jump higher), so research can cater to these qualities. Unlike sports, for music performance no **single optimum movement pattern can be generalized as a "best fit" for all performers** because differences in musical interpretation are not only acceptable, they are desirable. As a result, research in performance needs to differentially use any methodologies adopted from other disciplines. For researchers **it may be helpful to remember that "technique" in** performance is not the acquisition of idealized skills, rather it is the application of intention to a musical process; something that may be successfully accomplished using a variety of compensatory behaviors.

Compensatory behaviors, those movement strategies that are used to shift control between motor mechanisms, are frequently less important in sport than they are in music because most sport activities are dominated by gross motor control. In addition, for sports, anthropometry is often an exclusionary criterion (e.g. there are no 200 cm tall Olympic-level female gymnasts, just as there are no 150 cm tall male basketball players in the NBA). This makes anthropometry less of a factor in the development of training methods. In music performance, there are few exclusionary criteria. Anthropometry influences how the body interacts with the tools of performance, which is directly related to gestural timing and musical outcomes. What we perceive of as "differences in style" among performers may thus be partly due to anthropometric individualization.

In order to use tools of science to study music performance in meaningful detail, it becomes clear that new methodology must be developed. In the views of the authors, movement analysis technology provides significant potential in this regard because it has matured to a state where it allows quantitative evaluation of both gross and fine motor control. So far, the missing element for 3-D motion capture has been a biomechanical model that meets music's particular performance-related requirements. The current paper proposes, and in the discussion below demonstrates, the utility of a full-body, biomechanical model that considers: 1) gross and fine motor control, 2) their interaction, and 3) interplay that may be dependent on the tools of performance.

To demonstrate the potential of this method, results documented above serve as material for discussion of: 1) control pattern changes within single joints that compensate for playing conditions and/or the tools of performance, 2) compensatory interactions between joints, including gross to gross, fine to fine, and gross to fine control pairs, and 3) control pattern consistencies that may be linked to the effects of training.

Control pattern changes related to individual joints can be found for both gross and fine motor mechanisms. Data related to shoulder ab/adduction (Table 1) shows little change between the trials in terms of average joint angle (only 4.2°). However, averages do not tell the whole story. Shoulder (Ab/add) demonstrates notable change in control pattern for the two protocols that are furthest apart in terms of starting pitch (Figure 10, top row of graphs, dotted lines). These changes appear to coincide with points where strings crossings occur. Observed contour differences between the trials may be explained by the fact that the hand touches the upper bout of the violin (Figure 7) and must work around its curved shape when playing in higher positions (with the left hand closer to the violin body than the scroll). Postural compensation in gross motor control occurring between elbow and wrist (Figure 10, top row of graphs, dark and thin grey lines, respectively) may be explained as follows: the elbow angle must decrease to bring the hand into a higher position and the already bent wrist straightens somewhat to accommodate the shape of the violin. In addition, the general change in shape of the wrist graph seems to indicate behavioral changes from one that accommodates string crossings to one that does not; a phenomenon that may be explained by the fact that the wrist becomes nearly arrested when the hand is in constant contact with the violin bout.

Fine motor control changes can be exemplified by observing behaviors of the index and 4th fingers of the left hand. For the 3rd joint of index finger ROM, absolute angles and overall control pattern all change (Figure 11, middle row). ROM increases for the trial where the hand is touching the bout of the violin because the finger needs to reach across the body of the violin to touch the strings. For the same reason, the joint needs to be generally more open. The change in the control pattern may be necessitated by the two above-identified compensations. Observed control pattern changes in the 2nd joint of the index finger (although less dramatic than for the 3rd joint) may be attributed to sympathetic behavior throughout the finger's kinematic chain. This represents a strategy that involves two fine motor control variables (i.e. fine to fine compensation). For the first joint of the 4th finger, decreased ROM and a generally lower average joint angle in the higher position may indicate codependent relationship with the wrist (Figure 11, bottom row); if the wrist is more bent, the 4th finger is closer to the string and doesn't need to move as far to finger a note (gross to fine compensation).

Movement of the thumb, although clearly compensatory, is somewhat more difficult to categorize. Because we are used to thinking about the thumb as the "opposable" digit of the hand, it is natural to understand its behavior in terms of fine motor control in some instances and gross in others. In this case, its functionality mimics gross motor control more than fine. It compensates for (or perhaps anchors) all the other digits. Its behavior (among joints 1, 2, and 3) is comparable to that of the shoulder (among the components of Flex/ext and Ab/add). Motor control changes in 1st joint (Ab/add) of the thumb appear to coincide with strings crossings. Observed contour differences between the trials likely occur because the hand touches the upper bout of the violin (Figure 7) and the thumb cannot move as much (Figure 11, top row). General angle changes between 1st and 2nd joints of the thumb can be explained as follow: since the thumb anchors the hand to the neck of the violin, the mechanics of the hand reaching around the upper bout causes the thumb's 1st joint (Flex/ext) to decreases and the 2nd joint to increase. We believe that compensation in the thumb may also be linked to shoulder movement, an assertion that requires more detailed investigations.

A close examination of the bowing arm reveals bowing to be stable among trials. Consistency of joint control observed in Figure 10 (bottom row – right arm) clearly indicates that bowing did not accommodate starting position of the left arm and hand; rather it consistently accommodated string crossings. This indicates control stability due to years of training. Such training effectively eliminates sympathetic "cross-talk" between the left and right side functions of the arms. Since violin playing requires asymmetrical motor control and our protocol presented specific difficulties for the left hand, stability observed in the bow could only enhance the ability of the left hand to work without unnecessary interference. Intuitively, one might easily hypothesize that some elements must provide stability for successful performance for a contextual, forward-planned musical process that utilizes dynamic interchange of gross and fine motor control.

Finally, investigation of these kinds of compensatory strategies has potential for significance beyond the immediate goals of pedagogy. Finding ways to enhance learning, accelerating skill acquisition as well as increasing skill effectiveness, is a primary concern for both motor learning theory and music pedagogy. However, motor learning theory includes a third component – training to avoid injury (Ballreich and Baumann, 1996). In music practice, this has started to become more of a concern since overuse syndrome (OS) resulting from repetitive motion affects a significant percentage of performing musicians (50%-76%) (Zaza, 1998). String players are particularly affected with 48-66% of performers reporting injuries serious enough to interfere with their ability to play (Fry, 1986, 1988; Middlestadt and Fishbein, 1989). Although the current study documents data from a single individual it, in conjunction with the example of Adrian Anatawan, suggests that there may be many motor control options available to individual players. We speculate that performers' health and wellness may benefit from consciously invoking diversity in their compensatory behaviors during practice and performance rather than aiming to perfect a single idealized one. This hypothesis is supported by previous studies that show, compared to dynamic motor control patterns, static (or arrested) ones pose significantly higher risk for injury (G. Shan, Visentin, and Schultz, 2004; Visentin and Shan, 2003). By identifying alternative control patterns, compensation may become a preventative measure for controlling or reducing risk of injuries.

CONCLUSION

The goal of this study was to document a methodology that holds the potential to quantify gross and fine motor control behaviors in music performance. This case study of violin performance proved that 3-D motion capture combined with full body biomechanical modeling can reveal: 1) gross and fine motor control behaviors, 2) their interaction, and 3) effects that were dependent on the tools of performance. Further, the current paper substantiates the method`s broader utility through discussion of potential applications in pedagogy and injury prevention. The conceptual leap made in this paper is that there are many motor control means to achieve a desired outcome. In this ethos, compensatory strategies become tools for the performer rather than barriers to performance. Compensatory strategies are identified by examining: 1) control pattern changes within single joints, accounting for playing conditions and/or the tools of performance, 2) interactions between joints, including gross to gross, fine to fine, and gross to fine control pairs, and 3) control pattern consistencies that are linked to the effects of training. The range of research questions that may be answered by this methodology is exceedingly rich.

In summary, an understanding of the role of compensation as evocable strategy rather than a default condition has great potential to improve music pedagogy and vocational outcomes for performers. The method can be extended to all forms of instrumental performance where movement is involved. By linking interpretation of results to practical realities of the endeavor it is possible to provide pedagogical meaning that has applications across large cohorts of players as well as for individuals. Anthropometric differences may be a factor in individual style, but musico-cultural/aesthetic expectations fundamentally affect compensatory motor behaviors. The proposed method permits evaluation of the first of these. It remains to apply the method to a large number of subjects and musical contexts to begin to understand the second.

REFERENCES

Anantawan, A. (2012). Adrian Anantawan violinist Retrieved August 15, 2012, 2012, from http://www.adriananantawan.com/video.php

Ballreich, R., and Baumann, W. (1996). *Grundlagen der Biomechanik des Sports (The basics of biomechanics in sports)*. Stuttgart: Enke Verlag.

Buczek, F. L., Sinsel, E. W., Gloekler, D. S., Wimer, B. M., Warren, C. M., and Wu, J. Z. (2011). Kinematic performance of a six degree-of-freedom hand model (6DHand) for use in occupational biomechancics. *Journal of Biomechanics, 44*, 1805-1809.

Cerveri, P., Lopomo, N., Pedotti, A., and Ferrigno, G. (2005). Derivation of Centers and Axes of Rotation for Wrist and Fingers in a Hand Kinematic Model: Methods and Reliability Results. *Annals of Biomedical Engineering, 33*(3), 402-412. doi: 10.1007/s10439-005-1743-9.

Doidge, N. (2007). *The Brain That Changes Itself.* New York: Penguin Group.

Ferrario, V. F., Macri, C., Biffi, E., Pollice, P., and Sforza, C. (2007). Three-Dimensional Analysis of Hand and Finger Movements during Piano Playing. *Medical Problems of Performing Artists, 22*(1), 18-23.

Fry, H. (1986). Overuse syndrome of the upper limb in musicians. *Medical Journal of Australia, 144*(4), 182-185.

Fry, H. (1988). Patterns of over-use seen in 658 affected instrumental musicians. *International Journal of Music Education, 11*(1), 3-16.

Huron, D. (2006). *Sweet Anticipation: Music and the Psychology of Expectation.* Cambridge: Massachusetts: MIT Press.

Iltis, P., and Givens, M. (2005). EMG Characterization of Embouchure Muscle Activity: Reliability and Application to Embouchure Dystonia. [Research]. *Medical Problems of Performing Artists, 20*(1), 25-33.

Levetin, D. J. (2007). *This is Your Brain on Music: The Science of a Human Obsession.* New York: Plume.

Metcalf, C. D., Notley, S. V., Chappell, P. H., Burridge, J. H., and Yule, V. T. (2008). Validation and Application of a Computational Model for Wrist and Hand Movements Using Surface Markers. *IEEE Transactions on Biomedical Engineering, 55*(3), 1199-1209.

Middlestadt, S., and Fishbein, M. (1989). The prevelance of severe musculoskeletal problems among male and female symphony orchestra string players. *Medical Problems of Performing Artists, 4*(1), 41-48.

NIOSH. (1997). Musculoskeletal Disorders and Workplace Factors. *Publication No.*, 97-141.

Palmer, C. (2001). Movement amplitude and tempo change in piano performance. *Acoustical Society of America Journal, 115*(5), 2590-2590.

Schoonderwaldt, E., and Altenmüller, E. (2011). *Mastering the violin: Motor learning in complex bowing skills.* Paper presented at the International Symposium on Performance Science, Toronto, Canada. http://www.legacyweb.rcm.ac.uk/cache/fl0026872.pdf

Shan, G., and Bohn, C. (2003). Anthropometrical data and coefficients of regression related to gender and race. *Applied Ergonomics, 34*(4), 327-337.

Shan, G., and Visentin, P. (2003). A quantitative three-dimensional analysis of arm kinematics in violin performance. *Medical Problems of Performing Artists, 18*(1), 3-10.

Shan, G., Visentin, P., and Schultz, A. (2004). Multi-dimensional Signal Analysis as a means of better understanding factors associated with repetitive use in violin performance. *Medical Problems of Performing Artists, 19*(3), 129-140.

Shan, G., and Westerhoff, P. (2005). Full body kinematic characteristics of the maximal instep Soccer kick by male soccer players and parameters related to kick quality. *Sports Biomechanics 4*(1), 59-72.

Snyder, B. (2000). *Music and Memory An Introduction.* Cambridge: Massachusetts: MIT Press.

Turner-Stokes, L., and Reid, K. (1999). Three-dimensional motion analysis of upper limb movement in the bowing arm of string-playing musicians. *Clin Biomech, 14*(6), 426-433.

Visentin, P., and Shan, G. (2003). The kinetic characteristics of the bow arm during violin performance; an examination of internal loads as a function of tempo. *Medical Problems of Performing Artists, 18*(3), 91-98.

von Hasselbach, J. (2012). ***Für ‚lebendige' und ‚beziehungsfähige' Töne im*** *Streichinstrumentspiel. Eine Reaktion auf Peter Röbkes Einleitung zu Das Musizieren und die Gefühle.* Paper presented at the *Musik und Gefühl. Interdisziplinäre Annäherungen in musikpädagogischer Perspektive*, Hildesheim (Olms)/ Germany

Wikipedia. (2012, August 4, 2012). Adrian Anantawan Retrieved August 14, 2012, 2012, from http://en.wikipedia.org/wiki/Adrian_Anantawan#Media_coverage

Winter, D. (1990). *Biomechanics and Motor Control of Human Movement.* New York: Wiley.

Zaza, C. (1998). Playing-related musculoskeletal disorders in musicians: A systematic review of incidence and prevalence. *Canadian Medical Association Journal, 158*(8), 1019-1025.

In: Applications in Arts Biomechanics
Editors: Gongbing Shan and Peter Visentin
ISBN: 978-1-62808-015-5

Chapter 4

Effects of Training on Mass Balancing Oscillations in the Bowing of (Pre) Teen Violin Students: A Quantitative Micromotion Study

***Julia von Hasselbach*[*,1], *Wilfried Gruhn*[2] *and Albert Gollhofer*[1]**
[1]University of Freiburg, Germany
[2]University of Music Freiburg, Germany

Abstract

In a first study on mass balancing oscillations in the bowing of violinists (Hasselbach et al. 2010) the results supported the theoretically derived assumption that mass balancing oscillations constitute a characteristic pattern of expert violinists' motion. The question emerged as to whether it is possible to learn and train this ability within the scope of violin lessons in children. The present micromotion study uses 3D-kinematics and aims first to reveal the incidence of this characteristic pattern in the bowing of violinists between the ages of 10 and 15, and second to investigate to what extent a specially developed training affects mass balancing oscillations.

Two samples of ten string instrument players – with different teachers but matched in age and year of instrumental instruction – performed the same musical task as the adult university violin students in the 2010 study: repetitive motions in a G-major scale, but at different tempi. Mass balancing oscillations were measured and quantified as complementary motion of hand and elbow (CM). The incidence of mass balancing oscillations within the (pre) teen violin students was comparable to the lowest expert level in the 2010 study (see Figure 1). The difference between the mean values of the two samples was not significant and did not exceed 30% of systematic variance at the outset. One of the samples was investigated further in regard to possible training effects based on a repeated-measures design. The second sample served as a control group. In consequence of eight special exercises, the children tended to increase their mean value of CM significantly from 31% to 56% after fifteen weeks, whereas the control group

[*] Corresponding Author: Julia von Hasselbach, e-mail: julia@von-hasselbach.de.

increased the mean value of CM from 23% to 28%. Thus practice effects caused by the repeated measurement and maturation effects did not lead to significant differences.

Specific training of mass balancing oscillations as a supplement to musical training seems to be an advisable procedure in the instrumental instruction of children as early as the teenage years. It might enhance the chances of achieving higher expert levels. To show lasting effects, especially at higher tempi, training should probably continue over a longer period of time.

1. INTRODUCTION

Three-dimensional computer supported analysis of arm kinematics in violin players has been published since 2003 (Shan and Visentin 2003). Successive studies by Shan and Visentin aimed to provide data to reduce the high percentage of performing artists suffering from overuse syndromes. Biomechanical knowledge of playing string instruments has been expanded through empirical research since the beginning of the 20th century (see Steinhausen 1903/1928 and Trendelenburg 1925/1974). In particular, August Eichhorn, a solo cellist, university professor and researcher, succeeded in synthesizing his scientific and artistic ideas with prolific results (see Hopfer 1941 and Hasselbach et al. 2010). As revealed in depth in Hasselbach et al. 2010, Eichhorn analyzed the movements of his teacher, the famous cellist Emanuel Feuermann. He presumed that Feuermann was utilising the benefits of mass balancing oscillations to compensate for inertia, as presumably David Oistrach later did in the field of violin playing. Eichhorn proposed a complex theory of leverage in string instrumental playing corresponding to the sound-physical demands of string instruments. Playing with an economic leverage system involves the use of small force pulses and a dynamic change of mass positions. In recent decades empirical studies have also emphasized that expert violinists' motion is characterized by small muscle activity with appertaining ballistic effects (see Palac 1992 and Hasselbach 2009 for reviews). Mass balancing oscillations can be simply explained by imagining a seesaw and its mechanical characteristics. Some well-known pedagogues such as Katŏ Havas (Havas 1964/2003) or Gerhard Mantel (Mantel 1973/1976) have used this image. However, Eichhorn's theory of leverage is quite complex, with longitudinal and transverse seesaws working together in a mobile chain. The study of Hasselbach et al. 2010 was the first empirical support of Eichhorn's theory. (In fact, this theory of leverage is a reconstruction based upon a method which has been taught by one of Eichhorn's former cello students up to the present day: J. Schwab, at the 'Hanns Eisler' University of Music, Berlin).

Expertise is often connected with virtuosity, which is partly based on physical economy of physical execution, and musically convincing virtuosity depends on excellent sound quality (see Meyer 1978). The activity of balancing masses within the kinematic chain of the bowing arm and the application of ballistic effects in the interplay with the instrument should facilitate playing at faster tempi while maintaining good sound quality and a feeling of ease. (For detailed descriptions of significant relations between artistic sound differentiations and biomechanics, see Hasselbach et al. 2010). Instrumental instruction in teen violin students sometimes prepares for admission procedures at university or for participation in excellent youth orchestras. The achieved level of expertise is a crucial factor in a successful application, which often takes place for musicians at quite early ages. The 2010 study of

Hasselbach et al. analyzed the incidence of mass balancing oscillations in the bowing of adult university violin students. Mass balancing oscillations were measured and quantified as complementary motion of hand and elbow (CM). The students were divided into three expert levels according to their success in the admission procedure for a chosen degree as 1. school music teacher, 2. violin teacher, 3. solo performer (Figure 1).

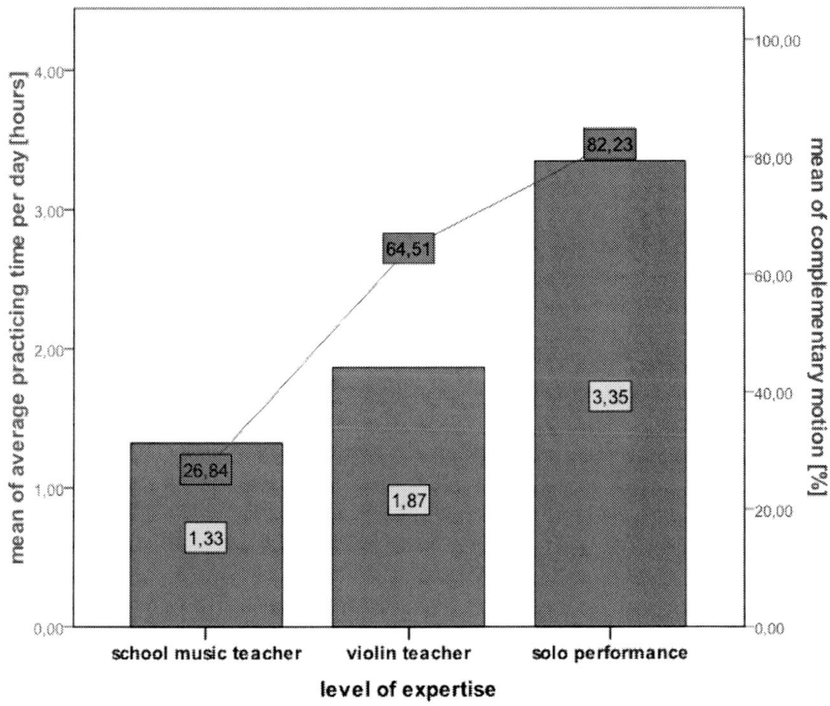

Figure 1. Comparison of the average practicing time per day with the percentage of complementary motion found at three levels of expertise in violin playing subjects, shown in Hasselbach et al. 2010.

The study revealed significant differences between the mean values of CM within the expert levels (one-way ANOVA with F=27.18 at the .05 level). Thus, a low level of expertise is suggested at CM values lower than 40%, a medium level of expertise at CM values between 40% and 70%, whereas a high level of expertise is characterized by CM values up to 95%. There is a lack of research on the incidence of mass balancing oscillations in the bowing of (pre) teen violin students. Additionally, the extent to which mass balancing oscillations can be stimulated by training has never been investigated. The purposes of this study are to reveal the incidence, and to evaluate the possibility of training mass balancing oscillations in violinists between the ages of 10 and 15. Because mass balancing oscillations within the use of an economic leverage system help to avoid static muscle tension in fingers, hand, arm, shoulder, neck and trunk, the results of this study may finally provide data for a later comparative analysis of healthy violin players and violin players suffering from overuse syndromes.

2. METHOD

Two samples consisting of ten secondary school children (2m/8f) each were compared cross-sectionally in their bowing movement in an experimental repeated-measures design. Students in both samples were in their third year of instrumental instruction, on average, but they had lessons with different teachers. One of the samples was investigated further in regard to possible training effects. The second sample served as a control group. Since the sample to be trained had already had up to three years of normal violin lessons with the first author, who instructed the performance of the exercises in this study, the possible mediating influences during previous violin lessons had to be controlled for effects on the CM values, especially due to an applied knowledge of biomechanics within the musical training. Thus, the obtained values of the control group additionally served as a cross-validation to check whether the subjects had already significantly higher values in complementary motion at the outset than other (pre) teen violin students. The children of the control group study at a municipal music school with a teacher who is less grounded in theoretical knowledge, but highly renowned as a violinist. As teachers change occasionally in the course of violin students' musical education, it was considered important to have a matched mixture of students in both samples who had either started playing violin with a certain teacher, or students who had joined a teacher after two years of violin playing in secondary school string classes. Additional data concerning the handedness of the subjects were collected via the Edinburgh Handedness Inventory. As the training measured motor and cognitive skills, additional tests concerning diverse aspects of intelligence seemed to be interesting. Therefore, the German version of the Wechsler Intelligence Scale for Children-IV ("HAWIK-IV") was applied to the trained subjects. The WISC-IV generates a Full Scale IQ (FSIQ) which represents overall cognitive ability. The four composite scores are Verbal Comprehension index (VCI), Perceptual Reasoning Index (PRI), Processing Speed Index (PSI) and Working Memory Index (WMI). All subjects and their parents gave informed consent.

Because of its simple application, without requiring complex calibrating, adjusting or programming, an easy transportable 'Zebris CMS 20' three-dimensional ultrasonic movement analysis system (Zebris Medical GmbH; Isny, Germany, Figure 2) was used to capture bowing movements on site during the violin lessons. The measurement method assesses the travel time of ultrasonic pulses, and guarantees high measuring accuracy (error for z-coordinate after triangulation < 1 mm in a distance of x < 1m). The movements in the musical task did not exceed a coverage radius of 1m. The coordinates were recorded with a rate of 65 measurements per second and marker. The miniature transmitters (markers) were placed on the distal ends of first and fourth metacarpal bones, that is, on the back of the hand in front of index finger (marker 1) and little finger (marker 2), and near the elbow on the lateral epicondyle (marker 3). Thus, at least one of the two hand markers transmitted the movement properly, if the other marker turned out of sight while playing on the G- or E-string. Additionally, the mean of the two hand marker values was taken into account for further calculation, in order to even out differences between the marker values due to possible slight transversal movements or trembling. The weight of these three markers (1g per marker) and cables did not considerably restrict the freedom of movement. The coordinates of the movements (triangulated raw signals) were gathered by the analysis software Win Data

(Zebris Medical GmbH; Isny, Germany) and exported into Microsoft Excel to calculate the percentage of complementary motion.

The children were asked to perform the same musical task as the adult university violin students had done in the preceding 2010 study. They had to play a two-octave G-major scale ascending and descending in pitch, with four repetitions of each note. Thus, 112 tones in a repetitive motion pattern could be investigated, in which one down-bow and one up-bow could form a more or less neat oscillation. The motions of elbow and hand can be complementary dephased in cases of mass balancing oscillations (see Figure 4), or run in parallel, in cases of inefficient activity of muscles working against the tendency to compensate for inertia (see Figure 3).

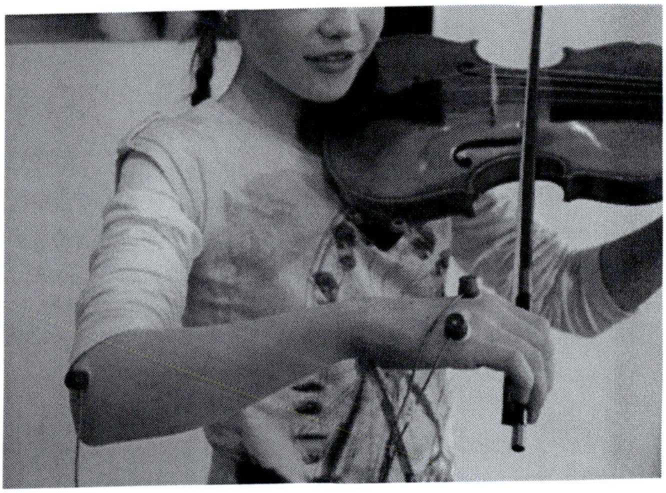

Figure 2. Violin student in the three-dimensional measurement of her bowing movement using the ultrasonic system Zebris CMS 20 (Zebris Medical, Isny, Germany).

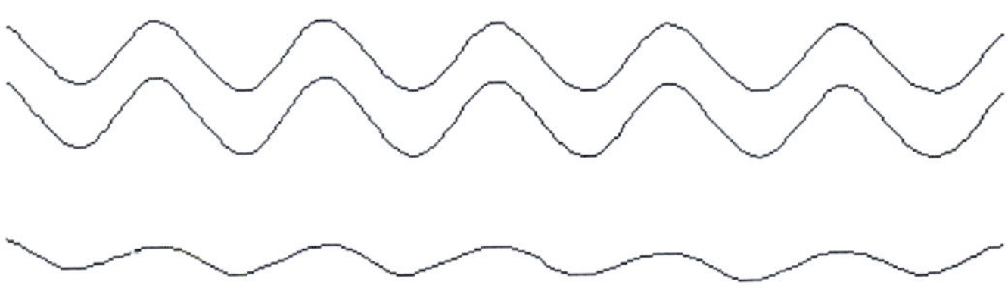

Figure 3. Parallel movement of elbow (lower line) and hand (upper two lines) from a subject with only 18% of calculated complementary motion.

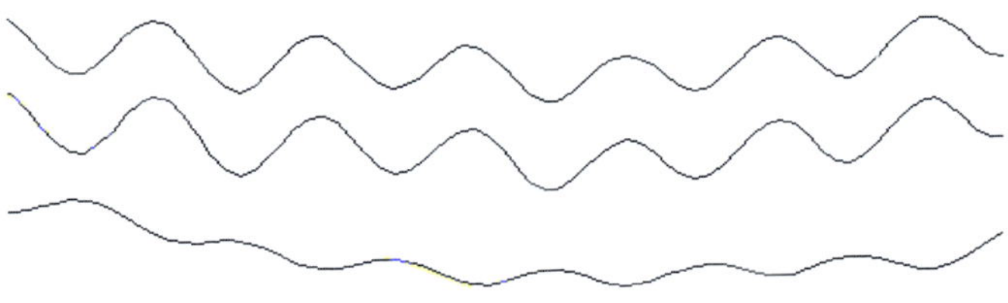

Figure 4. Dephased movement of elbow (lower line) and hand (upper two lines) from a subject with 85% of calculated complementary motion.

The musical task had to be played at three different tempi given by a metronome: first at about 30 M.M. per quarter note, second at about 60 M.M. per quarter note, third at about 80 M.M. per quarter note. But the children were free to perform the tempi very approximately, to avoid affecting the movement quality due to reduced concentration or a stiffening of the joints potentially triggered by metronome stress. Last, the task had to be played at a free fast tempo. The instruction was: "Play fast sixteenths with a good sound quality while feeling at ease – with a nice devil running after you!" The resulting tempi chosen differed widely, as did the tempi matched to the metronome pace as well. The mean values and the peak values of CM out of four trials for each subject were taken into account for statistical calculation. (For detailed information concerning the algorithm to quantify the movement quality, see Hasselbach et al. 2010.)

The children in the control group were aged 9.4 to 15.1 years (mean age 12.8) at their assessment (henceforth called "C"); they were aged 9.7 to 15.4 years (mean age 13.1) at the controlling assessment ("D"). The children in the sample to be trained were aged 10.4 to 14.4 years (mean age 12.1) at their first assessment (henceforth called "O" as in "outset") and they were aged 11 to 15 years (mean age 12.6) at the final assessment in February 2011 (called "B", whereas the intermediate assessment is "A"). The first measurement (O) took place shortly before the beginning of the 2010 summer holidays to avoid measuring after a longer interruption of practice, because the average practicing time per day was correlated with the percentage of CM in the previous study. The subjects in this study were also asked to divide their practice session times for the duration of the study into two parts, "practice time" of the study exercises and normal "musical practice". Both practice time and musical practice had to be logged.

The first period of systematic training took place in autumn, 2010. During 15 minutes within the scope of their normal violin lessons, the children learned how to practice eight basic exercises. Learning by imitating the teacher (average CM 79.37%, max. 92.41%, min. 69.83%; see Figure 5) was supplemented by explanations of the exercises, regarding the sound-physical and biomechanical basis of their musical necessity. The explanations intended to teach why and how the exercises relax muscles, why they require small force pulses, how they help to make use of gravity, weight balances and inertia, how they increase the awareness of sound-physical laws of the instrument, and finally, how the interplay of forces promotes good sound quality. After the subjects had learned the exercises for five weeks, an intermediate measurement of the dependant variable "complementary motion" took place (A). After this learning stage, an additional ten-week period of training and refinement followed (holidays excluded). For statistical analysis, three levels of training were defined as the independent variable in an ANOVA with repeated-measures: 1. before learning the exercises (O), 2. after five weeks of training (A) and 3. after fifteen weeks of training (B).

A description of the eight essential exercises of different types used for training follows below. The exercises were selected and/or developed to promote technical skills that advance sonority, especially mass balancing oscillations, since these are interrelated. To reduce complexity, only the longitudinal seesaws in the bowing arm were investigated in the measurement, although the exercises included training of all indispensable levers in the oscillating chain of balanced masses:

1. Détaché with mass balancing oscillations at the center third of the bow (middle)
2. Détaché with mass balancing oscillations at the lower third of the bow (nut)

3. Détaché with mass balancing oscillations at the higher third of the bow (tip)
4. Portati over the whole bow (ca. eight dots), powering mass balancing oscillations via forearm rotation
5. Roaming portati back and forth over the whole bow (quasi rhythmic pattern) while **rhythmically changing the direction of the arm's "macro"**-oscillation in accordance to the bowing direction (upward-moving elbow while down-bowing and downward-moving elbow while up-bowing)
6. Free drift in bowing (changing the contact height without losing sound quality by means of guiding the bow angle)
7. Relaxation of retaining muscles to make good contact with the string and then making swinging flights from nut to tip and vice versa, in order to produce sonorous tones (using effective bow and arm weight, i.e., gravity, inertia, and short force pulses)
8. Tensing and relaxing the left hand fingers with intermediate forearm rotation, to avoid straddling in the left hand and muscle tension in bowing due to co-innervations

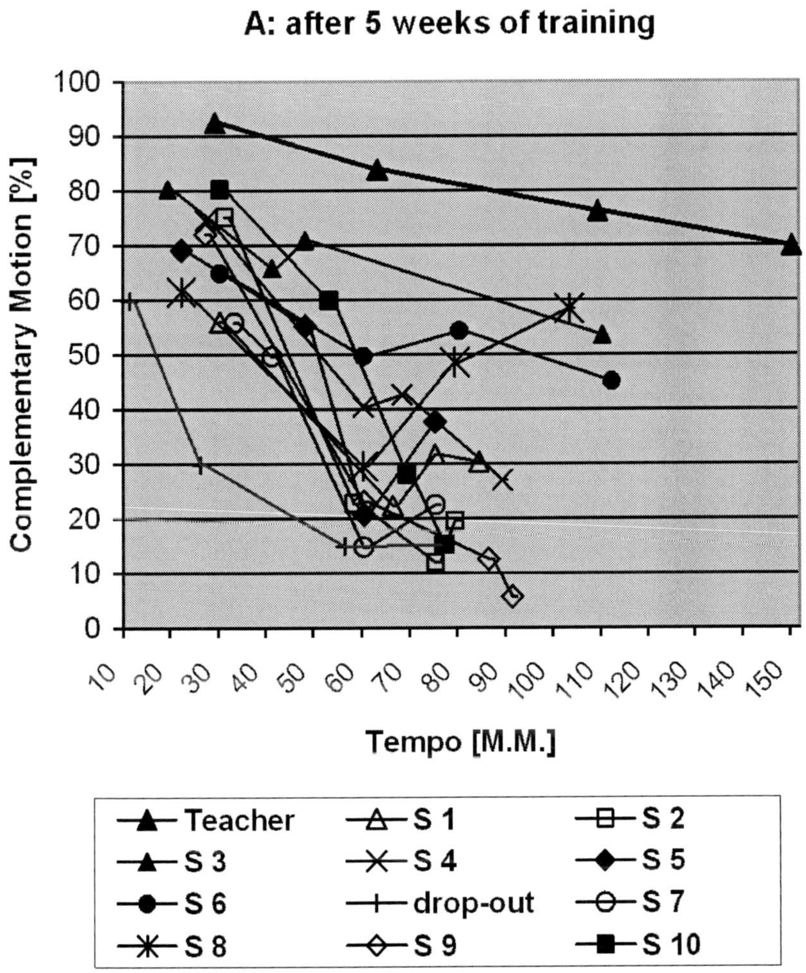

Figure 5. Visual graph of the data in measurement A.

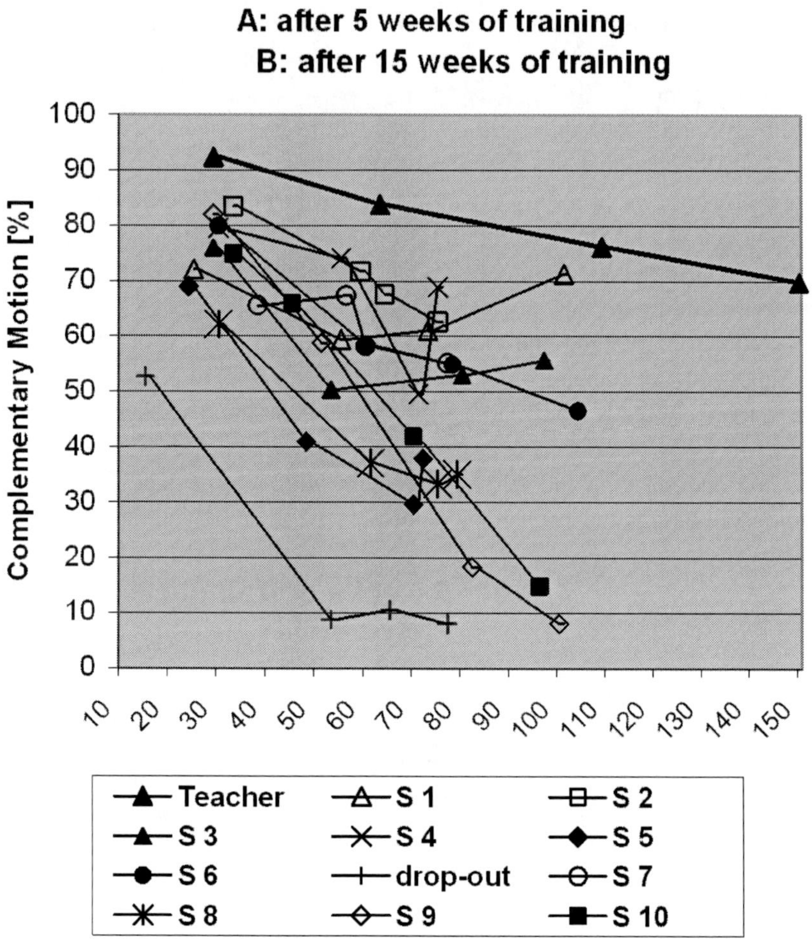

Figure 6. Visual graph of the data in measurement B.

The study started with twelve children, but after the first training period, one of the younger subjects quit the study due to a lack of motivation. A second subject was removed from the statistical analysis because of illness and a missing initial measurement.

3. Results

At the outset the control group was characterized by correlations (on the .05 level) between the year of instrumental instruction and 1. the age of the subjects (.643), 2. the average tempo (.665), and 3. the peak values of tempo played (.677). In contrast, the sample to be trained showed none of these correlations, but revealed correlations (on the .01 level) between the year of instrumental instruction and 1. the average CM (.931), and 2. the peak values of CM (.924). This indicates that the preceding years of instrumental instruction and perhaps the instruction methods of their teachers influenced the achieved CM values or the achieved tempi, respectively. Nevertheless, the mean value of average CM (regardless of the tempo played) within the sample to be trained (31.09%, max. 66.73%, min. 7.54%) did not differ significantly (mean difference: 8.15%, significance .277) from the mean value of CM within the control group (22.94%, max. 39.2%, min. 13.6%). The mean of peak values of CM within the sample to be trained was 39.21% (max. 76.14%, min. 10.03%) at the outset, and 30.89% (max. 49.08%, min. 17.59%) within the control group. In addition, the means of the peak values of CM did not differ significantly (mean difference: 8.32%, significance .33). As the purpose of the study was to detect considerable effects of training with regard to expert levels as mentioned above, effects of training could have a medium range of perhaps 10% to 30% up to a maximum of ca. 90% of CM. Therefore, one could accept these null hypotheses concerning the differences between the samples at the outset with a power of 90% for the small sample size (N = 20) provided the systematic variance exceeded 30% of the total variance ($\Omega^2 > 0.3$).

After the first training period A, the mean value of average CM within the experimental group increased to 43.98% (max. 67.61%, min. 28.32), and the mean of the peak values of CM to 68.93% (max. 80.27%, min. 55.78%). After the second training period B, the mean value of average CM increased to 56.33% (max. 68.05%, min. 41.83%), and the mean of the peak values of CM to 74.64% (max. 83.42%, min. 62.15%) (See Figure 8). As the Mauchly test was not significant, the sphericity was hypothesized, and it was possible to apply an ANOVA with repeated measures to investigate whether or not the mean values of CM within the three levels of training O, A and B differed. The probability for F = 8.439 was p = 0.003. Therefore, the effect of training on the dependent variable **"complementary motion"** is **significant at the .01 level. The factor "level of training" accounts for 48.4% of the variance** of complementary motion within the sample (effect size). Post hoc analysis (Bonferroni confidence interval) revealed that only the variance between the level O (outset) and the level of training B (fifteen weeks of training) was significant at the .05 level (mean difference B-O: 25.24), whereas the variance between the outset and the level of training A, as well as between A and B, might have been incidental (mean difference O-A: 12.89, A-B: 12.35; HSD: 14.95). Power analysis had a value of 93.1%.

For the control group, the mean values of CM did not differ significantly between the assessment and the controlling measurement after approximately fifteen weeks (mean of average CM: 27.53%, max. 54.60%, min. 11.18%; mean of peak values of CM: 37.46%, max. 66.37%, min. 19.00%; mean difference of average CM: 4.58%; mean difference of peak values of CM: 6.56%). The handedness within the control group (mean: 26.58, SD: 83.55) differed extremely from the handedness within the experimental group (mean: 88.20, SD: 16.10), but handedness did not correlate with the achieved CM values within both groups. Handedness showed only a very marginal negative correlation with the achieved CM values within the control group; therefore, it seems improbable that the left-handedness of several subjects in the control group caused the lower mean values within the control group, although the left-handed children played the violin in the right-handed position. In addition, gender had no significant correlations with the achieved CM values within both groups.

Considering the discrete measurements, there were no significant correlations between the CM values and the tempi played at the outset, neither for the experimental group nor for the control group. Only after training did correlation increase to .573 (A) and .529 (B) at the .01 level. The visual graphs of the data (see Figures 5 and 6) show that the CM values declined with increasing tempo. Figure 7 shows that some peak values of CM could be found at higher tempi within the control group, as well as within the study sample at the outset, whereas after training, the former peak values were reached from 'above' with increasing tempo.

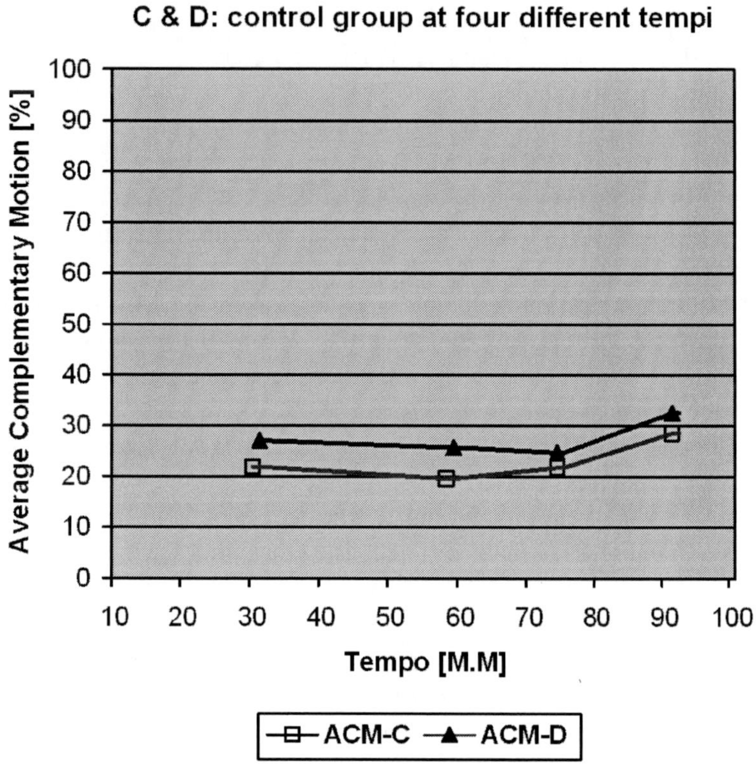

Figure 7. Visual graph of the data within the control group at the measurements C and D.

An astonishing result might be that neither the total practicing times during the training periods (including musical practice sessions) nor the separately gathered practice times of the exercises showed significant correlations with the achieved amount of complementary motion – neither with the average nor with the peak values of CM (mainly achieved at the slow tempi), but only with the tempi played. At the level of training A there were correlations of .916 between practice times and peak values of tempo (.01 level), of .781 between total practicing times and average tempo (.01 level), and of .677 between practice times and average tempo (.05 level). At the level of training B there were correlations of .71 between practice times (in B) and average tempo, as well as .71 between the accumulated practice times (in A and B) and average tempo (.05 level).

After the first training period, neither the year of instrumental instruction nor the age of the subjects had significant correlations with the achieved CM values, but after the second training period, the subjects' ages correlated with the achieved average CM (.725 at the .05 level). Cognitive skills might have influenced the achievements. Most of the subjects within the trained sample revealed outstanding IQ indices (mean: 123.7, min: 108, max: 139, SD: 8.79). There were no correlations between the CM values and the Full Scale IQ or the indices of the composite scores at the outset. After training, the achieved peak values of CM correlated significantly with the Working Memory Index (.632 at the .05 level). Surprisingly, the average CM values revealed significant but negative correlations with the Perceptual Reasoning Index (-.827 at the .01 level) and with the Processing Speed Index (-.666 at the .05 level).

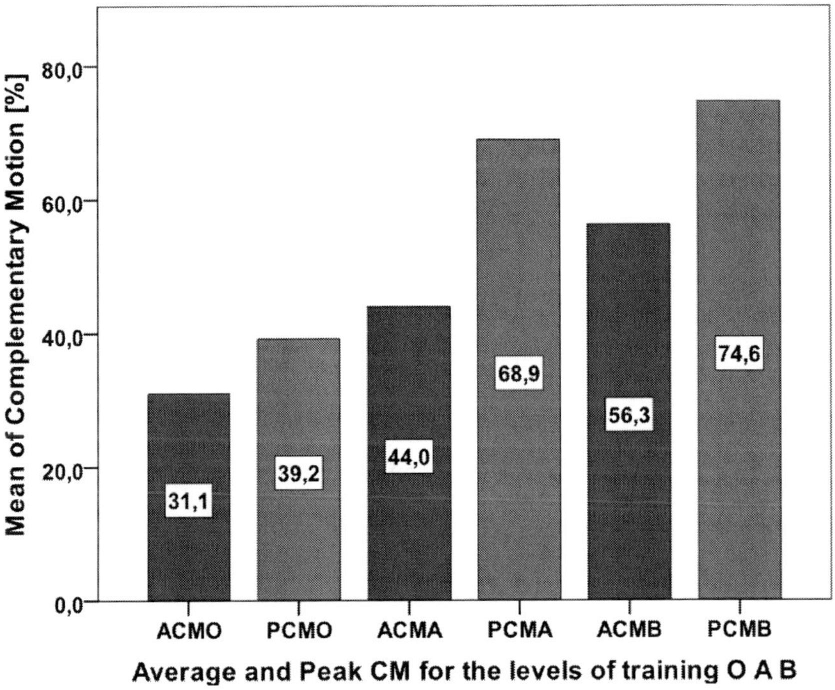

Figure 8. Means of average and peak values of complementary motion at the training levels O (Outset), A (after five weeks of training) and B (after fifteen weeks of training).

DISCUSSION AND IMPLICATIONS

The average incidence of mass balancing oscillations in the bowing of (pre) teen violin students – as far as they were achieved within the study samples – lies at a low level of expertise, comparable to the lowest expert level within the adult violin students (mean of CM<40%; for reference values in the bowing of adult violin students at different levels of expertise, see Figure 1). Only two of the subjects (20%) to be trained showed average CM values at a medium level of expertise higher than 60%, and peak values of CM higher than 70% (up to 76.1%) at the outset of the study. Within the control group, only one (10%) of the subjects had average CM values higher than 40%, but 50% of the subjects had peak values of CM higher than 40%. After the second training period, all trained subjects had average CM values higher than 40%, and all peak values were higher than 60%, up to 83.4 %, which lies within the range of high expertise. But the peak values of CM were not found at the peak tempi. The Figures 5 and 6 show the tendency in the teacher model and the similar tendencies in the subjects' data. When comparing this tendency with the correlation of .886 (at the .05 level) between the tempo played and the respective CM values in the 2010 study, in which the subjects with the higher CM values played at higher tempi, one can see an obvious inconsistency. One explanation could be that a decreasing tendency can be found within individual subjects, probably because more time at a slower tempo helps them control and feel the motion needed to produce the desired sound-quality, whereas accuracy decreases with increasing tempo. But for a sample of subjects, who did not play the task at fixed different tempi, the statistical result of higher CM values at higher tempi can be seen in relation to the respective individual level of expertise. The extent to which mass balancing oscillations are used may influence the feeling at ease and thus the tempo chosen and correctly played. In addition, both decreasing tendency and a change of the motion pattern applied at higher tempi (as suggested by Shan and Visentin) may occur within individual subjects, depending on how consciously the movement has been trained.

In addition, the 2010 study showed a high correlation between the subjects' accumulated practice times per day and the developed amount of CM (.896 at the .05 level, see Figure 1). This could not be confirmed directly by the results of this study. The amount of short-term practice time seems not to be a primary factor for the quality of bowing movements with respect to mass balancing oscillations in trained children, because mass balancing oscillations can be made consciously at slow tempi. But long-term practicing is important for the process of automating mass balancing oscillations, in order to have them at one's disposal at higher tempi. This may be confirmed by the finding that the trained children tended to increase the extent of mass balancing oscillations within 5 weeks to an average CM of 44%; however, the effects turned out to be significant only after the second training period of ten weeks, due to a stabilization of the movement quality at higher tempi. Thus, the mean values of CM from four trials at different tempi increased to an average CM of 56.3%. (See figures 5 and 6) To show lasting effects, especially at higher tempi, training should probably continue over a longer period of time.

Within the control group, effects of practice caused by the repeated measurement and maturation effects did not lead to significant differences, but a tendency toward increased values after fifteen weeks may be indicative of a slow development over time in normal violin lessons. A sudden improvement occurred in one out of the ten subjects. This improvement accounts for nearly the whole difference between the mean values of the measurements C and D. The distribution of CM values within the adult university violin students suggests that CM values can remain stable at a lower expert level over the whole span of instrumental instruction. The intra-individual development of CM values with increasing tempo makes a pattern strikingly stable, as it was within the mean values of the whole group (see Figure 7). The teacher of the violin students in the control group as the imitated model may be the causal factor for this pattern. The teacher did not want to have his CM values measured, but this characteristic pattern of a change at higher tempi was estimated by sight and suggests an interrelation.

The results of the WISC-IV intelligence indices are plausible in the positive correlation of the peak values of CM with the Working Memory Index causing an intense processing at slow tempi. The negative correlations of the average CM with the Perceptual Reasoning Index and the Processing Speed Index are unexpected and oppositional to previous assumptions. It may be that the high scores at these indices are associated with a tendency to overly control the movements, which can be counterproductive at higher tempi.

The study suggests that mass balancing oscillations can be learned and trained at an early stage of instrumental instruction with (pre) teen violin students. Specific training of mass balancing oscillations as a supplement to musical training seems to be an advisable procedure in violin lessons. It might promote the artistic development and enhance the chances of admission to violin studies at higher university levels. As the effects of training correlated not with the practice times, but with the age of the subjects, training with teenaged children may be more effective than training with preteen children (see the data visual graphs in Figures 5 and 6, with the results arranged according to declining ages of the subjects). It may be that the crucial factor for successful training is not only the ability to intellectually understand the theoretical backgrounds of the exercises, but the emotional openness to reap the musical benefits of good sound production while using mass balancing oscillations. Especially with very young children, motivation to train as rigorously as was done within the study might be problematic. The younger children seemed to participate only because of a good relationship with the teacher, but without personal enthusiasm, whereas some older students seemed to realize the importance and the musical benefits of the training. Especially when struggling with technical difficulties while learning musical pieces, these students benefitted from the transfer of the trained movements which led to sudden insights and to a desire for further related success in playing the violin.

Although the ill subject was left out of the statistical analysis, it is nevertheless interesting to look at this subject's CM values. As shown in Figures 5 and 6, the CM values were extremely low. Significantly, this subject suffered from a spinal blockage. This may indicate the relevance of mass balancing oscillations for a healthy approach to playing. Furthermore, it underlines the necessity of measuring the movement qualities, because the teacher did not completely realise the scope of the 'bad' movements by sight alone during the training period. Training of mass balancing oscillations holds some risks if not done carefully with an experienced instructor whose listening is constantly tuned for the resulting sound quality. Research on the effects of the training approach with regard to artistic development or to medical aspects must be undertaken in further studies.

References

Hasselbach, J. von (2009). 100 Jahre Physiologic Turn in der Streichinstrumentalpädagogik. Eine Bestandsaufnahme. In Schläbitz, N. (ed.), *Interdisziplinarität als Herausforderung musikpädagogischer Forschung* (pp. 239-262). Essen: Blaue Eule.

Hasselbach, J. von and Gruhn, W. and Gollhofer, A. (2010). Mass balancing oscillations. An indication of expertise in the bowing of violinists. A quantitative micromotion study. In *Proceedings of the 11th ICMPC*. Seattle: University of Washington.

Havas, K. (1964/2003). The twelve lesson course. London: Bosworth.

Hopfer, M. (1941). Die Klanggestaltung auf Streichinstrumenten. Das Naturgesetz der Tonansprache; kurze Einführung in "Die gestaltende Dynamik der Bogenmechanik" von August Eichhorn. Leipzig: Kistner and Siegel.

Mantel, G. (1976). Cello Technique. Forms and Principles of Movement. Indiana: University Press; German Version (1973). Cello Technik. Bewegungsprinzipien und Bewegungsformen. Köln: Gerig.

Meyer, J. (1978). Physikalische Aspekte des Geigenspiels. Ein Beitrag zur modernen Spieltechnik u. Klanggestaltung für Berufsmusiker, Amateure u. Schüler. Siegburg: Schmitt.

Palac, J. A. (1992). Violin bowing technique. An analysis of contemporary pedagogical literature according to principles of human movement. *Medical Problems of Performing Artists*, 7, 30–34.

Shan, G., and Visentin, P. (2003). A quantitative three-dimensional analysis of arm kinematics in violin performance. *Medical Problems of Performing Artists*, 18, 3–10.

Shan, G. and Visentin, P. and Schultz, A. (2004). Multidimensional signal analysis as a means of better understanding factors associated with repetitive use in violin performance. *Medical Problems of Performing Artists*, 19, 129–139.

Steinhausen, F. A. and Reuter, F. von (1903/1928). Die Physiologie der Bogenführung auf den Streich-Instrumenten. 5. edition. Breitkopf and Härtel. Leipzig.

Trendelenburg, W. (1925/1974). Die natürlichen Grundlagen der Kunst des Streichinstrumentspiels. Unchanged photomechanical reprint of the edition 1925. Kassel: Hamecher.

Visentin, P. and Shan, G. (2003). The Kinetic Characteristics of the Bow Arm During Violin Performance. *MedicalProblems of Performing Artists*, 18, 91–97.

II – Dance

In: Applications in Arts Biomechanics
Editors: Gongbing Shan and Peter Visentin
ISBN: 978-1-62808-015-5

Chapter 5

TURNOUT IS AN EULER ANGLE

J. Shippen*

Coventry University, Coventry, UK

ABSTRACT

Euler angles are a widely accepted methodology for the description of joint articulation. This paper demonstrates that the third Euler rotation, in accordance with ISB co-ordinate definitions, represents the turnout of the lower limb which is of critical significance in dance.

This novel interpretation of the Euler angle was found to be practical, mathematically rigorous and, in contrast to current methods, applicable to any dance posture for the calculation of turnout. Additionally, it can be applied to calculate the constituent elements of total turnout attributable to turnout at the hip and ankle together with tibial twisting motion.

The technique was applied to a sample of dancers performing demi and grand pliés in first, second, fourth and fifth positions. The mean and standard deviation of the turnouts throughout these movements are presented enabling the identification of atypical turnout patterns which will assist screening and the identification of potential injury risks.

Keywords: Dance, turnout, Euler angles

INTRODUCTION

Correct turnout of the lower extremities is the most important anatomical factor in classical ballet and its function is to provide balance, agility, speed and strength essential to performance. To achieve turnout, the legs are externally rotated from the hips although a perfect turnout of 90^0 at the feet is anatomically rare. Forcing the feet to achieve a turnout position beyond which the hips will allow can aesthetically distort the position of the entire

* Corresponding author: Dr James Shippen, Department of Industrial Design. Coventry University, Coventry CV1 5FB. The United Kingdom, Telephone: +44 – 24 – 7688 – 7072, Email: j.shippen@coventry.ac.uk.

body (Gilbert et al, 1998) and cause injuries (Clippinger, 2005). Moreover, it has been suggested that the encouragement of ineffective and possibly damaging compensatory strategies to achieve the perfect turnout can lead to musculoskeletal injuries (Ryan and Stephens, 1987).

The majority of ballet movements use turnout at the beginning and end of movements and during passing from one movement to another (Bussell, 2000). It is therefore necessary to **measure a dancer's turnout during the whole of a dynamic dance** movement, rather than merely for the initial posture, to provide objective analysis of turnout to improve performance and reduce the risk of injuries.

Until recently dance was dominated by qualitative descriptive studies and has only lately embraced the advantages which quantitative methods can provide (Grossman et al, 2005). It is agreed that turnout results from summative contributions of the hip, knee, lower leg and the foot-ankle complex. However the most frequently reported measurement is hip external rotation whilst the dancer is static and in an initial pose although even this is reported in incompatible ways (Champion and Chatfield, 2008). Turnout is currently assessed using a variety of techniques which makes comparison of results from different studies difficult. Assessment techniques include, amongst others, subjects standing in first position on a protractor (Bennell et al, 1999), tracing feet onto paper and measuring the angles formed by lines drawn on the paper (Watkins et al, 1989) or photographs of dancers standing on friction-free disks (Meinel and Atwater, 1988). Orthopaedic assessment of turnout is often obtained by passively measuring hip external rotation (Clippinger-Robertson, 1990) by means of an orthopaedic examination, for example, using a goniometer whilst the dancer lies in a prone position, with hips in extension, knee flexed to 90° whilst the hip joint is passively externally **rotated by the examiner moving the dancers' hip** joint through its range of motion. This reflects a measurement protocol recommended by the American Academy of Orthopaedic Surgeons and approved by international associations of orthopaedic surgeons (Surgeons, 1965).

However, none of the above methodologies are applicable to measure dynamic choreographed dance movements. Moreover, projection based methods become irrelevant if projecting onto a non-near orthogonal plane. For example, the projection onto the floor of a femeral anterior/posterior axis is reasonable in the initial turnout stance where turnout is normally measured, however the projection is not reasonable in stances where non-orthogonality is large. This situation will occur during significant knee flexion in the grand plié and many other steps.

The International Society of Biomechanics (ISB) recommends co-ordinate systems associated with each of the major anatomical segments, for example the pelvis, the foot, the shank (Wu et al 2002; Wu et al 2005). The ISB also recommend the Joint Coordinate System (JCS) for the description of joint articulations (e.g. hip extension/flexion, abduction/adduction, internal/external rotation). However this method of describing joint movements is identical in concept to an Euler angle based description. The JCS joint articulations are described as three rotations about axes fixed to the proximal and distal segments and about a third axes which is perpendicular to the segmental fixed axes. In the ISB recommendation, the distal segment axis of rotation, the e3 rotation, is about the long axis of the segment which therefore aligns with the turnout rotation. Therefore in this study, Euler angles between adjacent segments of the lower limbs are calculated and the e3 rotation is identified to be a measure of turnout.

This paper will calculate the Euler angles between adjacent segments throughout dance movements in order to observe the turnout. The method provides a robust 3-dimensional definition of the turnout angle measurement for any posture, both statically and dynamically and is suitable for practice, rehearsal or performance. Such information on turnout cannot be derived from other approaches currently commonly employed to evaluate dance movements.

METHODS

The purpose of this study was to apply the ISB recommended joint articulation definitions and relate that definition to a generalised methodology for calculating ballet turnout consisting of hip joint rotation, tibial twisting motion and rotation at the ankle whilst the lower extremity is in any posture and during dynamic movements.

In accordance with the ISB definitions, each lower extremity was considered to be composed of three segments; the femur, the lower leg and the foot. The pelvis was also considered to be a segment. The turnout of the lower extremity was defined to consist of contributions of rotations between these segments and hence will be referred to as 'segmental turnout'. The mutual rotation of these segments can be unambiguously calculated using a variety of methods including Euler angles or Joint Coordinate System (JCS), helical screw angles and Cardan angles (Kaplan, 1976). In accordance with ISB recommendations, a co-ordinate system was defined on each of these segments as described by Wu et al (2002) (Figure 1) and the JCS was used to define the joint articulation.

The Euler/JCS approach specifies the angles which rotate a proximal segmental reference frame to the orientation of a distal reference frame (see Appendix). Euler angles specify rotations about a moving segment reference frame; i.e. the rotation is applied about the axes of a co-ordinate system which rotates with the distal segment. The ISB recommendation for the sequence of rotations at the hip is:

e1-flexion / extension
e2-abduction / adduction
e3-internal / external rotation of the femur

The rotations at the knee are:
e1-knee flexion
e2-varus / valgus
e3-tibial twisting motion

and the rotations at the ankle are:

e1-dorsiflexion / plantaflexion
e2-inversion / eversion
e3-internal / external rotation

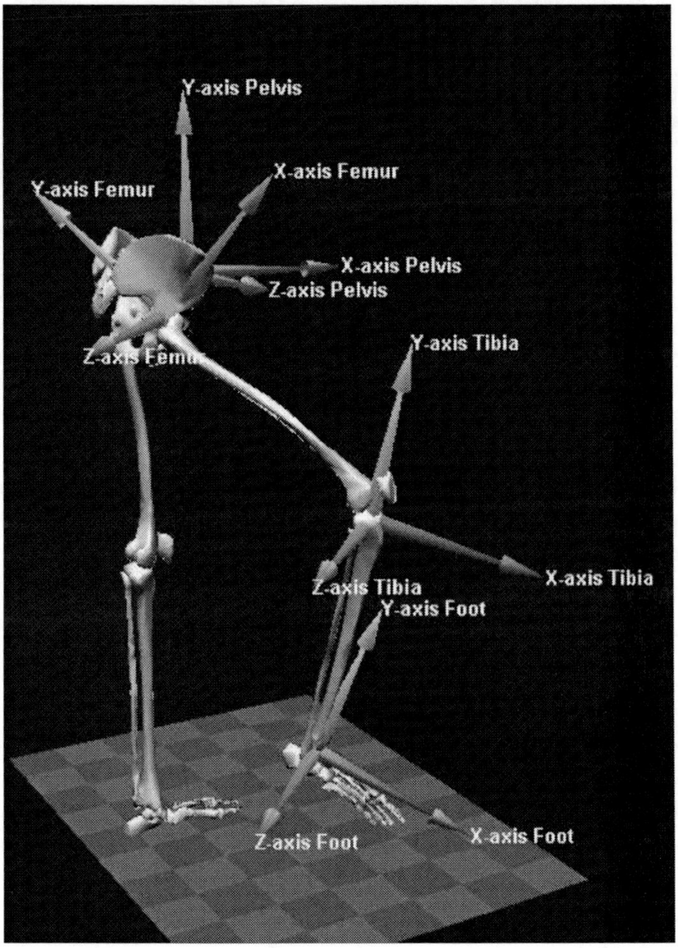

Figure 1. Illustration of ISB recommended co-ordinate definitions.

The third rotation for each of the above joints is associated with the turnout rotation of the lower limb. Therefore if the Euler angles are calculated for these joint articulations and the e3 rotation is isolated, this will be the turnout at each joint.

Additionally, the angle between the foot and the sagittal plane (FTSP) was calculated. This angle is defined as the angle between the projection of pelvis x-axis onto the floor and the projection of the foot x-axis onto the floor. Hence this angle is an approximation of the traditional turnout angle defined at the feet.

The above Euler angle based approach was used to measure the segmental turnout of ten female professional ballet dancers (hence 20 legs) aged 18 to 28 years who had at least ten years ballet training and no current injuries. The subjects gave informed consent and the study was approved by the University Research Ethics Committee.

The movement of the dancers was measured using a 12 camera MX40 Vicon 3-dimensional optical tracking system. Retro-reflective markers were attached to the dancers in a full-body marker set similar to the Helen Hayes marker set (Collins, Ghoussayni, Ewins and Kent, 2009) but the medial markers on the knee and ankle were retained throughout the trial. The movement was sampled at 50 frames per second. The data was filtered using a 4th order low-pass digital Butterworth filter with a 20Hz cut-off frequency implemented in Matlab™.

From the movement of the retro-reflective markers, and hence the anatomical landmarks on the twenty lower extremities of the dancers, the orientations of all of the segmental co-ordinate systems were calculated. The data was processed in the SAM biomechanical analysis environment. The only defining landmark to which a marker could not be attached was the hip joint centre. Therefore the location of the hip joint centre was calculated using an interpolation method from the locations of the right and left anterior superior iliac spine markers and the right and left posterior iliac spine markers (Bell et al, 1989).

Prior to the trials, the subjects were asked to stand in a neutral posture and the transformation matrices were calculated from the reference frames in the idealised stance to the reference frames in the measured stance; i.e. the calibration transformation. When the orientation of the recorded reference frames was measured during the trials, they were firstly transformed by the inverse of the calibration transformation. This reduced the affect of misalignment of the reference frames due to marker misplacement. It also accounted for the anteversion angle of the tibia and excluded this permanent set from the calculation of the tibial torsional angle.

After an initial warm-up period, each dancer was requested to perform two demi pliés and two grand pliés in first, second, fourth and fifth positions without the barre. First position is the posture where the heels are touching. The second position is the posture where the ankles are separated by approximately the width of the pelvis. In the fourth position the heel of the forward foot is aligned with the mid foot of the rear foot, the forward foot is approximately one foot length forward of the rear foot. In fifth position both feet are touching, the toes of each foot reaches the heel of the other. In all positions, the feet are externally rotated.

Demi and grand pliés were selected as this involves large rotations at the hip, knee and ankle and orientations of the lower extremity which is far from orthogonal to the floor; a situation where traditional turnout methods are unsuitable. The subjects were asked to identify which of the two demi and grand pliés in each position they considered to be indicative of best practice and therefore to be analysed. The subjects were not informed that the study was pertaining to turnout nor were the subjects instructed to maximise turnout.

The movements of each of the subjects were measured using the optical tracking system throughout the demi and grand plies.

From this positional data the orientation of each of the anatomical segments was calculated. The Euler angles of joint articulations were then calculated in accordance with the ISB recommendations. The rotation of the distal segment about its long axis, the e3 rotation, which represents turnout was recorded.

To compare the turnout angles throughout the movement, the time base of each trial was replaced with a metric which represented the fractional completion of the task. The maximum flexion in demi plié was normalised to 0.33, the maximum flexion in grand plié was normalised to 0.67.

The end of the trial was normalised to 1. This common independent variable (i.e. fractional completion of the trial) enabled the calculation of the mean of the hip turnout, tibial twisting motion, ankle turnout and FTSP. Symmetry conditions were used to calculate the mean and standard deviations across the right and left sides of the body resulting in a sample size of 20 lower extremities.

RESULTS

Figure 2 shows the mean e3 turnout angles for each leg as the dancers performed the demi plié and grand plié in the first position. The standard deviation of the hip turnout angle, turnout due to tibial twisting motion, ankle turnout and FTSP for the dancers in the first position are illustrated in figure 2 as grey shading surrounding each of the mean plots.

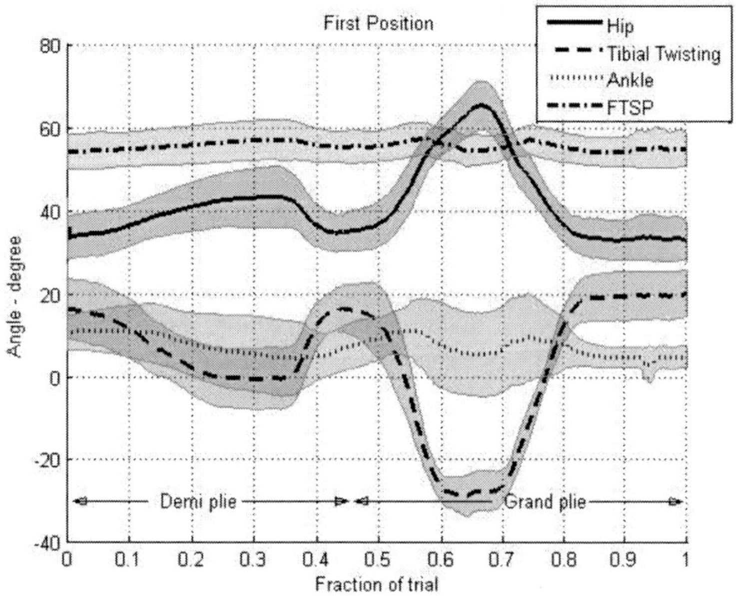

Figure 2. The mean segmental turnout with a ±1 standard deviation envelope during demi and grand plié in first position.

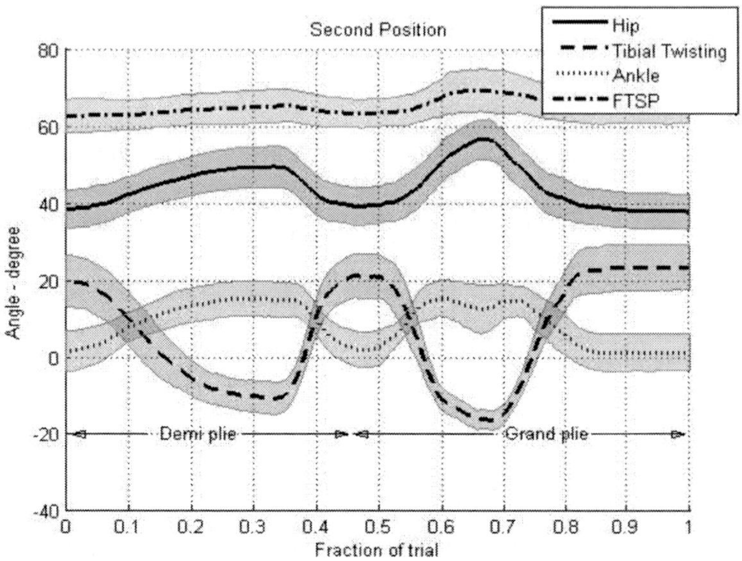

Figure 3. The mean segmental turnout with a ±1 standard deviation envelope during demi and grand plié in second position.

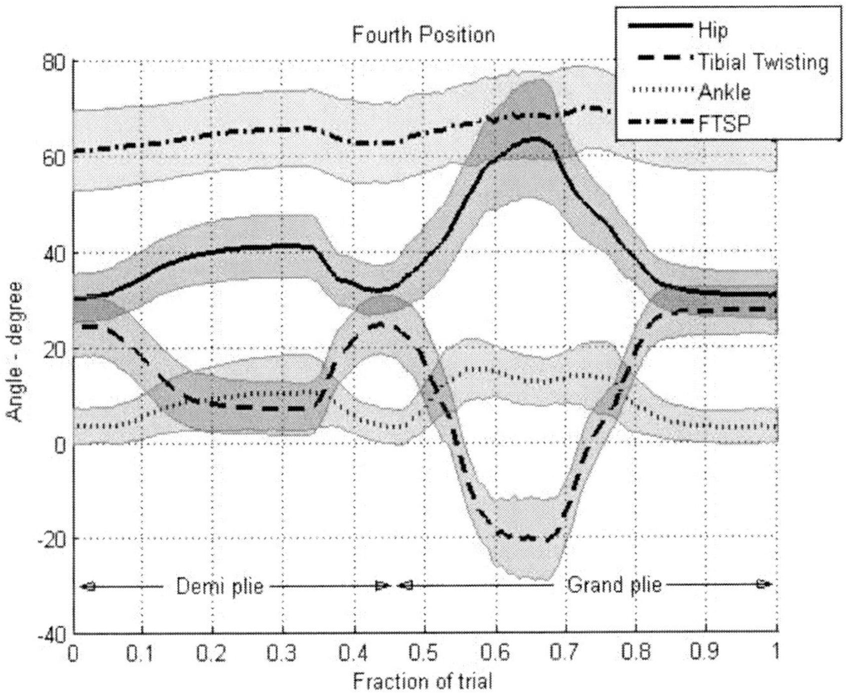

Figure 4. The mean segmental turnout with a ±1 standard deviation envelope during demi and grand plié in fourth position.

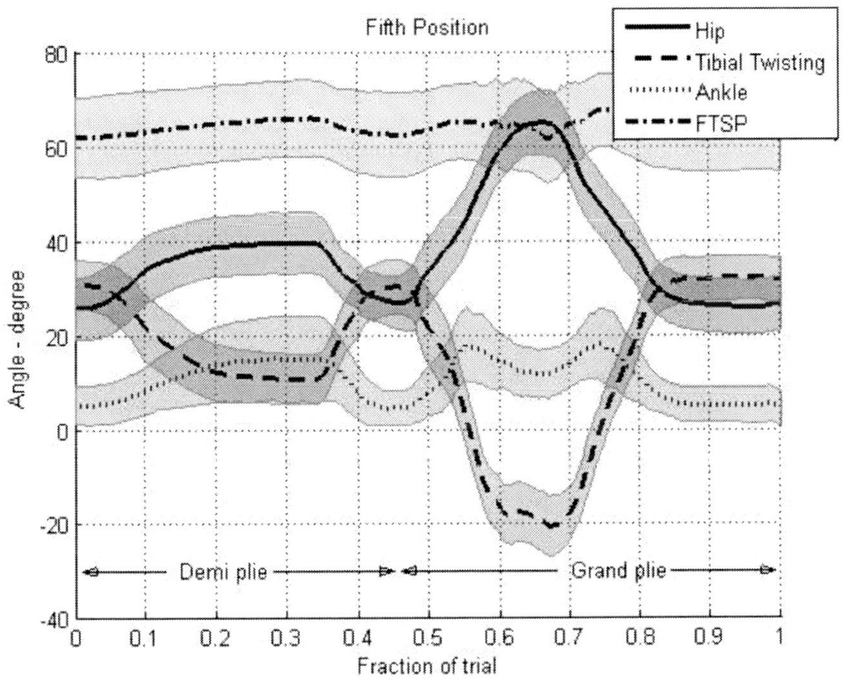

Figure 5. The mean segmental turnout with a ±1 standard deviation envelope during demi and grand plié in fifth position.

The width of the shaded area represents ±1 standard deviation about the mean for the sample of dancers. The plots illustrate turnout from the initial stance and throughout the maintenance of turnout during a dynamic movement. It is seen that there are only small changes in the hip, tibial twisting motion and ankle turnout (except tibial twisting motion in demi plié for the second position) until the maximum flexion of grand plié is approached. At the point of maximum flexion, the changes in hip turnout were comparable to the change in tibial twisting motion. For all positions in grand plié, at maximum knee flexion there is observed the increase in turnout at the hip and a retrograde movement in the tibial twisting motion. This consistent with increasing knee flexion increases rotation capacity at the hip and knee.

It should be noted that the sum of the segmental rotations approximates to the FTSP angle only in the initial posture of minimal hip and knee articulation. As a counter example, Figure 1 illustrates the lower extremity in an arbitrary position where the turnout angles are not aligned and hence these angles cannot be simply summed.

DISCUSSION

As Grossman (2003) has indicated dancers are frequently uncertain how much turnout they can achieve. This may be due to the current situation where different techniques for the measurement of hip external rotation produce quite different results (Clippinger 1990) and hence dance teachers are unable to objectively assess the degree to which their students are expressing turnout.

Despite this lack of a unified, rigorous consensus on a protocol for the measurement of turnout, the literature indicates that of the desired 90° of perfect unilateral external rotation, approximately 60% of turnout is generated at the hip (Grossman, 2003; Liederbach, 1997). Additionally, researchers (Bennell et al 1999; Negus et al, 2005; Martin et al, 1998; Kahn et al, 2000) have reported mean first position turnout measurements ranging from 93° to 131° depending on dancers' age and experience and the measurement protocol used. The results from this study are in general agreement with these disparate results, however all of these other measurements relate only to the dancers in the initial posture whereas this study has extended the measurement to a dynamic scenario.

The diversity of results from the literature indicates that there is a need for a standard procedure to be able to accurately measure all the constituent parts of turnout i.e. hip rotation, tibial twisting motion and ankle rotation (Gilbert et al, 1998). Moreover the effectiveness of undertaking measurements during movement has been recommended based on the assumption that such measurements will be beneficial not only to clinicians, teachers, and trainers but also to dancers themselves (Watkins et al 1989; Molnar, 1995) to identify their individual maximum turnout capabilities.

Therefore a mathematically rigorous, 3-dimensional methodology based on the calculation of Euler angles between adjacent segments of the lower extremity has been proposed for the calculation of turnout along the limb. This approach is a novel interpretation of the ISB recommended method for the measurement of joint articulations. This is not limited to a static, initial posture as with traditional approaches but can accommodate any dynamic choreographed movement.

As an application of the segmental turnout method, the technique was applied to dancers performing a demi plié followed by grand plié. It was found that there were small changes in turnout during demi plié and large changes of turnout, both at the hip and tibial twisting motion, around maximal knee flexion of grand plié. This is consistent with the unlocking of the tibial twisting motion occurring during knee flexion. Throughout, ankle turnout was small.

However, the application of this method requires knowledge of the 3-dimensional orientation of each of the lower extremity segments together with the pelvis. To undertake this measurement during movement requires access to a 3-dimensional tracking system. In practicality, this limits the application of the method to dance biomechanics research institutions. However this approach greatly extends the situations under which turnout can be measured. Much of the research into injury associated with incorrect turnout has been undertaken in the initial static position due to an inability to measure turnout in other stances. Using a Euler angle based approach these studies can be extended to study turnout over an entire movement.

CONCLUSION

A new interpretation of Euler angles has been presented which express the turnout of dancers in generalised, dynamic postures. This approach follows the recommendation of the ISB on the measurement and reporting of joint articulations. The results in terms of plots of turnout against a metric which represented the fractional completion of the task (i.e. real, or normalised time) can be easily understood by dance students, dance teachers and clinicians. Additionally, the total turnout, FTSP, has been decomposed into its constituent parts of turnout at the hip, ankle and tibial twisting motion. This segmental approach to the measurement of turnout has been found to be practical to apply. The means and standard deviations of turnout for the sample during demi and grand pliés in first, second, fourth and fifth positions are also presented. The small standard deviation in comparison to the calculated turnout angles across the cohort of dancers suggests the robust nature of this approach as a turnout descriptor. An individual dancer whose turnout characteristics are remote from the mean may be susceptible to injury although this will require further research.

ACKNOWLEDGMENT

The author would like to thank the Leverhulme Trust for supporting this research.

REFERENCES

Bell, A., Petersen, D. and Brand, R. (1989) 'Prediction of the Hip Joint Centre Location from External Markers.' *Human Movement Science* 8, 3-16.

Bennell, K., Khan, K. and Matthews, B. (1999) 'Hip and Ankle Range of Motion and Hip Muscle Strength in Young Female Ballet Dancers and Controls.' *British Journal of Sports Medicine* 33, (5) 340-346.

Bussell, D. (2000) *The Ballet Book*. London: Dorling Kindersley Ltd.

Champion, L. and Chatfield, S. (2008) 'Measurement of Turnout in Dance Research: A Critical Review.' *International Association of Dance Medicine and Science* 12, (4) 121-135.

Clippinger, K. (2005) 'Biomechanical Considerations in Turnout.' In *Preventing Dance Injuries*. 2nd edn. ed. by Solomon, R. Champaign, Ill.: Human Kinetics: 135-150.

Clippinger-Robertson, K. (ed.) (1990) Biomechanical Considerations in Turnout. New York: *J. Association of the American Alliance for Health*, Physical Education, Recreation and Dance.

Gilbert, C., Gross, M. and Klug, K. (1998) 'Relationship between Hip External Rotation and Turnout Angle for the Five Classical Ballet Positions.' *J. Orthop. Sports Phy Ther*. 27, (5) 339-347.

Collins, T.D., Ghoussayni, S.N., Ewins, D.J., and Kent, J.A. (2009). A six degrees-of-freedom marker set for gait analysis: repeatability and comparison with a modified Helen Hayes set.' *Gait Posture* 30(2):173-80.

Grossman, G. (2003) 'Measuring Dancer's Active and Passive Turnout.' *Journal of Dance Medicine and Science* 7, (2) 49-55.

Grossman, G., Krasnow, D. and Welsh, T. (2005) 'Effective Use of Turnout: Biomecanical, Neuromuscular and Behavioural Considerations.' *Journal of Dance Education* 5, (1) 15-27.

Kahn, K., Bennell, K. and Ng, S. (2000) 'Can 16-18 Year Old Elite Ballet Dancers Improve Their Hip and Ankle Range of Motion over a 12 Month Period?' *Clin. J. Sports Med.* 10, 98 -103

Kaplan, M. (1976) *Modern Spacecraft Dynamics and Control*. New York, NY: John Wiley and Sons

Liederbach, M. (1997) 'Screening for Functional Capacity in Dancers: Designing Standardized, Dance Specific, Injury Prevention Screening Tools.' *J. Dance Medicine and Science* 1, (3) 93 -106.

Martin, S., Marquez, F. and Ordonio, M. (1998) 'Measuring Functional Range of Motion in Ballet Dancers' Hips.' *J. Dance Medicine and Science* 33, 340-346.

Meinel, K. and Atwater, A. (1988) 'Analysis of Components of the "Turnout" in Beginning and Advanced Female Ballet Dancers.' *Med. Sci Sports Exerc*. 20, (Suppl 2, S-6).

Molnar, M. (1995) 'Dynamic Evaluation of the Student Dancer; *Enhancement of Healthy Functioning and Movement Quality*.' Impulse 3, 287-295.

Negus, V., Hopper, D. and Briffa, N. (2005) 'Associations between Turnout and Lower Extremity Injuries in Classical Ballet Dancers.' *J. Orthop. Sports Phys. Ther*. 35, (5) 307-318

Ryan, A. and Stephens, R. (1987) *Dance Medicine a Comprehensive Guide*. Chicago Il.: Pluribus Press Inc.

Surgeons, American Academy of Orthopaedic Surgeons (1965) *Joint Motion: Method of Measuring and Recording*. Chicago IL.

Watkins, A., Woodhull-Mcneal, A. and Clarkson, P. (1989) 'Lower Extremity Alignment and Injury in Yhoung Pre-Professional, College and Professional Ballet Dancers, Part 1, Turnout and Knee-Foot Alignment.' *Med. Probl Perform Artists* 148-153.

Wu G, Siegler S, Allard P, Kirtley C, Leardini A, Rosenbaum D, Whittle M, D'Lim D, Cristofolini L, Witte H, Schmid O, Stokes I (2002) 'ISB recommendation on the definitions of joint coordinate system of various joints for the reporting of human joint motion - Part 1: ankle, hip and spine' *J. Biomech.* 35 543 – 548.

Wu G, van der Helm F, Veeger H, Makhsous M, Van Roy P, Anglin C, Nagels J, Karduna A, McQuade K, Wang X, Werner F, Bucholz V (2005) 'ISB recommendations on the definitions of joint co-ordinate systems of various joints for the reporting of human joint motion – Part II: shoulder, elbow, wrist and hand' *J. Biomech.* 38 981 – 992.

In: Applications in Arts Biomechanics
Editors: Gongbing Shan and Peter Visentin

ISBN: 978-1-62808-015-5

Chapter 6

SINGLE LEG POSTURAL SWAY CHARACTERISTICS OF DANCERS DURING A ROTATING TASK: A PILOT STUDY

Nicholas Caplan [*] ***and Alan St Clair Gibson***
School of Life Science, Northumbria University, Newcastle upon Tyne, UK

ABSTRACT

The aim of this study was to investigate the single leg sway characteristics of dancers whilst balancing on a rotating platform that was being used as part of a specific choreography they were training to perform. Seven professional dancers held a single leg standing posture on a rotating platform for as long as they could. The rotating platform was mounted on a force platform allowing the measurement of sway variability in the antero-posterior and medio-lateral directions. Participants' sway was analysed over the initial and final 10 seconds of the trial. Medio-lateral sway was found to be similar to antero-posterior sway at the start ($p = 0.766$) and end ($p = 0.938$) of the trial. Six of the seven participants showed reductions in both AP and ML sway. Although no statistically significant changes in antero-posterior ($p = 0.201$) or medio-lateral ($p = 0.064$) sway were observed between the start and end of the trial with all seven participants included, reductions in sway were seen that showed large effects (AP = 0.94; ML = 1.55). With only the six participants included in the analysis that showed reductions in sway, ML sway was found to significantly reduce between the start and end of the trial ($p<0.05$). This reduction in postural sway over time suggests that postural feedback could be initially perturbed, returning towards baseline levels over the course of the trial.

Keywords: Balance, rotation, sway, standing, centre of pressure

[*] Correspondence to: Dr N Caplan, School of Life Sciences, Northumbria University, Northumberland Building, Newcastle upon Tyne, NE1 8ST, United Kingdom, Telephone: +44 (0)191 243 7382, Email: nick.caplan@northumbria.ac.uk

INTRODUCTION

Many everyday human movements involve rotational actions, which puts increased demands on the balance systems of the body. During quiet stance, the visual, vestibular and somatosensory systems are able to satisfactorily maintain balance. However, if the body is rotated about a vertical axis, both the visual and vestibular systems will be perturbed, placing a greater "demand" on the somatosensory system (Caplan, Johnson, Horsburgh, and Walker, 2009). This is especially true in dancers during dynamic motions, who are required to move their eyes and head for artistic purposes, as well as to determine their orientation in the performance space. Golomer et al. (1999) found that professional dancers were less vision dependent than untrained subjects. It was suggested that professional dance training shifts the dominant feedback strategy from the visual system to the somatosensory system.

In order to investigate the role of the visual, vestibular and somatosensory systems in the control of posture and balance, studies have used either bipedal or single leg stance tests. Studies looking at single leg stance on a stable base of support have used centre of pressure measures to show changes in antero-posterior and medio-lateral postural sways (Hertel, Olmsted-Kramer, and Challis, 2006; O'Connell, George, and Stock, 1998; Suponitsky, Verbitsky, Peled, and Mizrahi, 2008; Verhagen, et al., 2005). Hertel *et al.* (2006) found reduced variability in sway in the medio-lateral direction compared to sway in the antero-posterior direction. Similarly, O'Connell *et al.* (1998) found slightly reduced ranges of sway in the medio-lateral direction for their control group. As part of a novel dance choreography, a group of professional contemporary dancers were required to perform for up to an hour whilst on a rotating platform that caused their bodies to rotate about a vertical axis. The platform rotations would act to remove the stable base of support normally used to maintain upright posture (Hugel, Cadopi, Kohler, and Perrin, 1999), thus reducing the effectiveness of the balance systems. The ability of the dancers to overcome this reduced effectiveness of the balance systems is an important consideration in the design of rotating choreographies. In order to examine the extent to which the use of a rotating platform for a prolonged period influences dancers' ability to maintain a stable posture, this study aimed to determine the influence of whole body rotation about a vertical axis on postural sway characteristics during a single leg balance task.

METHODS

Participants

Seven professional dancers (4 female) participated in the study, with a mean (±SD) height, mass and age of 1.66 (±0.06) m, 60.7 (±9.2) kg and 32 (±7) years, respectively. The dancers were recruited from a contemporary dance company.

At the time of the study, the dancers were rehearsing for at least four hours each day, for five days a week. Ethical approval was obtained from the institutional ethics committee, and the dancers provided written informed consent to take part in the study. The dancers were all using rotating platforms during their rehearsals at the time of the study, and had had one day of familiarisation to the rotational nature of the choreography.

Equipment

A platform was used to generate the required rotation of the participants about a vertical axis. The platform was provided by the dance company, and consisted of a wooden circular horizontal surface with a diameter of 1 m and a height of 0.2 m. The rotating platform was controlled by a motor (MDEMAXX 071-32, Lenze, Bedford, UK) to allow angular velocities of up to 1.2 rad.s^{-1}. The rotating platform was located centrally on top of a force platform (OR6, AMTI, MA) (Figure 1) that was located in the centre of a large room, with at least 8 metres distance between the platform and the walls. The forces and moments in and about the three orthogonal axes were amplified (gains: F_x=4000; F_y =4000; F_z =996.1; M_x =4000; M_y =4000; M_z=4000) by a strain gauge amplifier (SGA6-4CE, AMTI, MA) before being sampled at 1000 Hz by an analogue-digital board (DT3002, Data Translation, Inc., MA) and stored on a PC for later analysis.

Protocol

The dancers were asked to maintain a static single leg standing posture, with their metatarsophalangeal joints of the supporting foot positioned centrally on the platform (Figure 1). The heel of the non-supporting leg was placed on the knee of the supporting leg, which was slightly flexed, and both hands were placed on the non-supporting knee. The arms were kept relaxed, and the trunk and head remained upright and still in relation to the rotating base of support throughout the trial.

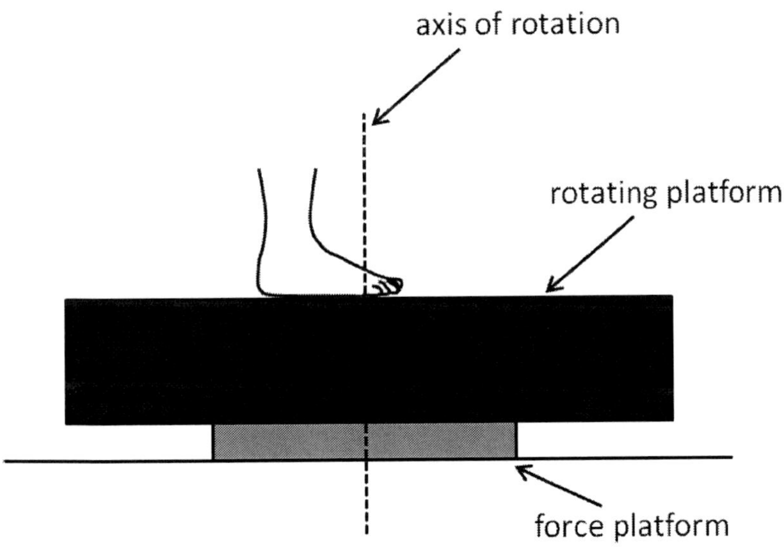

Figure 1. Side view of the rotating platform, showing its position on top of the floor mounted force platform. The position of the foot is also shown, illustrating the off centre location of the foot when the metatarsophalangeal joint was placed over the axis of rotation.

This position, which was one of the postures achieved during the choreography, ensured that the positions of the non-supporting leg and hands were controlled between dancers. The dancers were instructed to maintain a stable posture for as long as they could, , whilst the base of support rotated their bodies about a vertical axis at a constant angular velocity of 0.38 $rad.s^{-1}$. This angular velocity was the same that the dancers used for their choreography. Data collection started at the instant the platform rotations commenced. Each participant completed a single trial.

Data Processing

Due to the rotating platform being mounted on top of the force platform, the x- and y- centre of pressure displacements measured by the force platform did not correspond directly to the antero-posterior and medio-lateral directions of the participants. It was therefore necessary to decompose the x- and y- centre of pressure oscillations measured by the force platform (global coordinate system - GCS) into the antero-posterior (x) and medio-lateral (y) axes in the local coordinate system (LCS) of the platform. Due to the fully enclosed design of the rotating platform, it was not possible to mount a potentiometer on its vertical axis to directly measure the angular orientation of the platform. It was, therefore, not possible to measure directly the angular orientation of the platform in relation to the GCS. Thus, it was necessary to determine this angular orientation from an assessment of the signals measured.

Due to the position of the supporting foot relative to the centre of the platform, there was an offset of the centre of pressure oscillations away from the centre of rotation. This generated sinusoidal changes in the x and y centre of pressure data measured by the force platform in the GCS. As the heel was furthest from the centre of rotation, it was known that the anterior sway was directed towards the centre of the platform and that the medio-lateral sways were tangential to the circumference of the circular motion. However, it was not possible to decompose the centre of pressure patterns relative to the GCS into antero-posterior and medio-lateral sway (relative to the LCS of the platform) without first determining the angular orientation of the platform over time.

To determine the angular orientation of the platform over the duration of each trial, a fast Fourier transform was first applied to the x-axis (GCS) centre of pressure signal. From the resultant fundamental frequency, the angular velocity of the platform rotations was determined. The platform was also filmed at 50 Hz using a miniDV video camera (HVR-A1E, Sony, Japan) in order to determine the time elapsed for 5 revolutions to check the validity of the spectral analysis method. A 3rd order low pass Butterworth filter was then applied to the x-axis (GCS) centre of pressure data with the cut off frequency set to just greater in magnitude than the rotational frequency of the platform (cutoff = 0.1 Hz). This provided a signal containing only the low frequency oscillations of the centre of pressure data derived from the platform rotation. The times of each maxima and minima were found and the time scale adjusted so the first maxima corresponded to $t = 0$. Each maxima corresponded to the point of maximum posterior sway (i.e. away from the centre of platform rotation).

A 2nd order band pass Butterworth filter was applied to the raw centre of pressure signals in both the x- and y- directions of the GCS to remove the low frequency noise component of the signal due to the platform rotations and high frequency mechanical noise from the platform motor. The lower cut off frequency was equal to that used above to isolate the low frequency noise component of the signal and the upper cut off frequency was set to 25Hz as suggested by Morrison et al. (2007).

The antero-posterior, $COP_{A\text{-}P}$, and medio-lateral, $COP_{M\text{-}L}$, components of postural sway, relative to the LCS of the rotating platform, were then decomposed from the x- and y- centre of pressure signals, relative to the GCS, using the equations,

$$COP_{A\text{-}P} = COP_y \sin\theta - COP_x \cos\theta \qquad (1)$$

$$COP_{M\text{-}L} = COP_y \cos\theta - COP_x \sin\theta \qquad (2)$$

respectively, where COP_y and COP_x are the centre of pressure signals measured with respect to the two horizontal axes of the GCS, θ is the angular orientation of the rotating platform.

The standard deviations of both antero-posterior and medio-lateral sways were then calculated to give a measure of sway variability, as reported previously for single leg sways (Hertel, et al., 2006), for the first and last 10 second periods of each trial. A paired samples t tests was used to determine the significance of any difference found between antero-posterior and medio-lateral sway, and to determine any differences in postural sway between the start and end of the trial. A 95% confidence interval was used. Effect sizes (mean standard deviation of the two groups divided by change in mean between groups) were also calculated according to the method of Cohen (1988) where 0.2 indicates a small effect size, 0.5 is a moderate size, and 0.8 is a large effect size.

RESULTS

Participants were able to maintain the required posture during the rotational task for 66 ±32 seconds. Figure 2 shows the centre of pressure displacements of an example data set relative to the x and y axes of the force platform (GCS). The circular pattern generated by the positional offset of the centre of mass of the participant from the centre of the rotating platform can be clearly seen. This offset allowed for the sway signals to be decomposed into AP and ML sway in relation to the rotating platform.

Spectral analysis of these data showed that the rotations of the platform were at a frequency of 0.061 Hz (Figure 3A), equating to a platform angular velocity of 0.38 $rad.s^{-1}$. This was also confirmed from the video footage. Figure 3B illustrates the sinusoidal changes in the raw centre of pressure location in the x-axis (GCS), as well as the filtered x-axis (GCS) centre of pressure signal containing only the low frequency signal of the platform rotations. The maxima and minima indicate the points at which the anterio-posterior (LCS) direction and x-axis of the force plate (GCS) corresponded to each other. The time offset between $t = 0$ and the time of the first maxima for the example shown (Figure 3B) was 7.81 s.

Figure 4 shows the sway patterns (centre of pressure) now expressed relative to the antero-posterior and medio-lateral directions (LCS of the rotating platform). The circular pattern resulting from the positional offset between the centre of rotation and the centre of mass is no longer apparent. For the example participant shown, AP and ML sway was similar, with ranges of approximately 0.05m being seen in both directions.

Six of the seven subjects showed reductions in both AP (Figure 5A) and ML (Figure 5B) sway between the start and end of the trial. One participant showed a slight increase in AP and ML sway between the start and end of the trial.

The mean standard deviations of antero-posterior (AP) and medio-lateral (ML) sway for all participants are shown in Figure 5. Sway variability was similar between the AP and ML directions at both the start ($t = .311$, $df = 6$, $p = 0.766$) and end ($t = .081$, $df = 6$, $p = 0.938$) of the balance trial. When all seven participants were included in the statistical analysis, no significant changes in sway variability were seen between the start and end of the trial for antero-posterior sway ($t = 1.435$, $df = 6$, $p = 0.201$), although a large effect was seen ($ES = 0.94$)(Figure 6A).

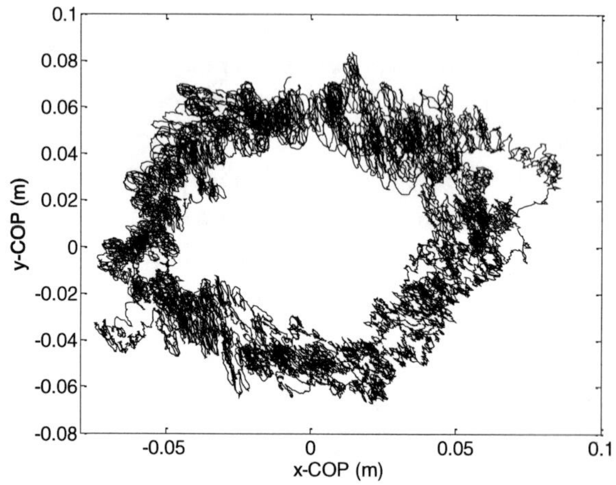

Figure 2. An example sway pattern on the rotating platform. The data is expressed relative to the global coordinate system of the force platform. The rotational influence of the platform can be seen due to the offset between the platform centre and the location of the centre of pressure within the base of support.

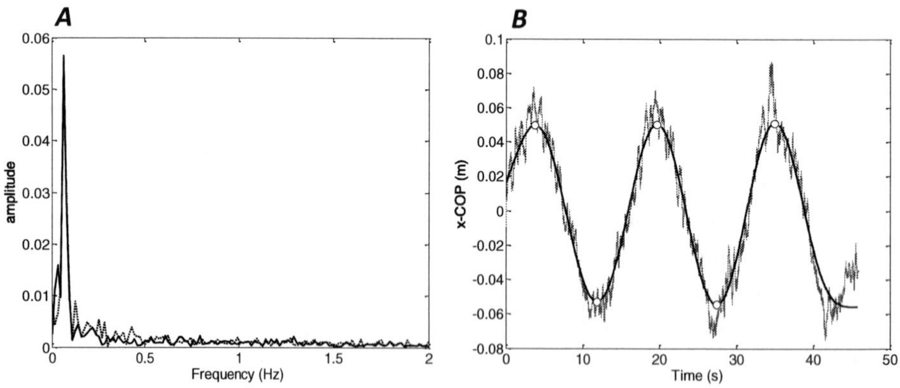

Figure 3. (A) Power spectrum of the centre of pressures in the x- (—) and y- (---) axes of the global coordinate system. The dominant low frequency signal indicates the angular frequency of the platform rotations. (B) Raw and filtered centre of pressures are shown for sways in the x-direction of the global coordinate system. The turning points are illustrated (o) from which the duration of each platform rotation were determined.

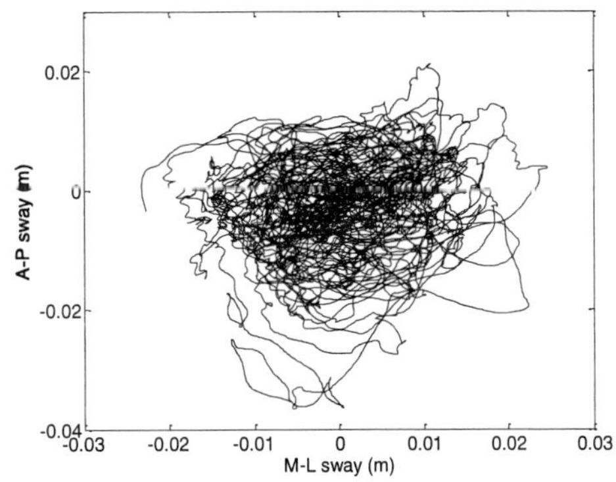

Figure 4. Sway patterns are shown relative to the anterio-posterior and medio-lateral axes of the local coordinate system of the rotating platform.

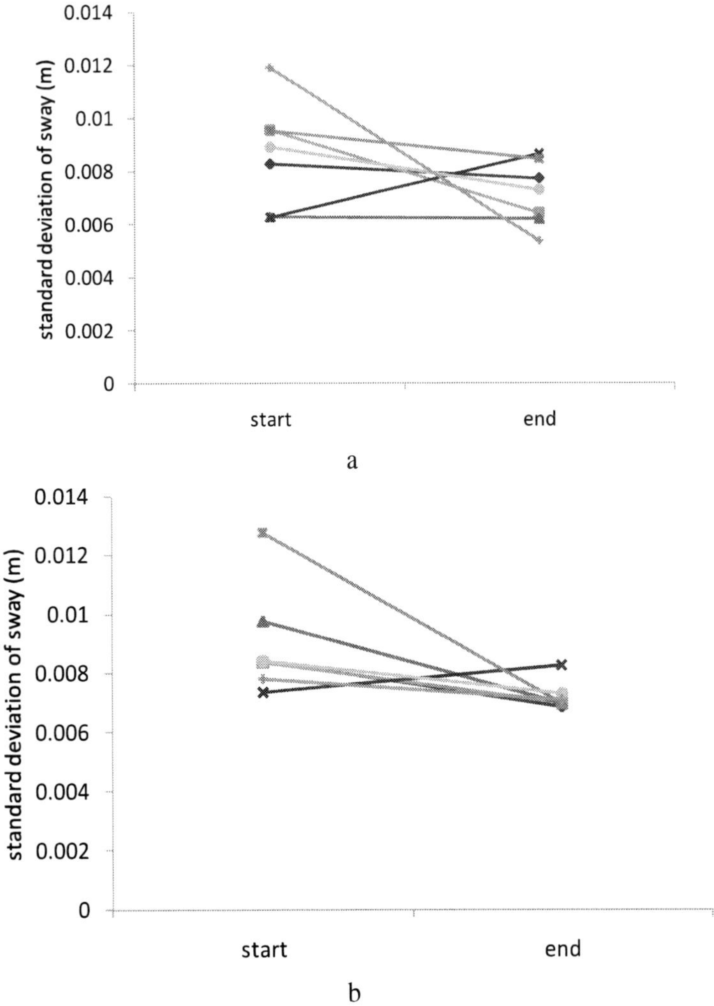

Figure 5. Standard deviations of sway are shown for each individual participant during the first and last ten second periods of their trial for AP (A) and ML (B) directions.

If the subject who showed an increase in AP sway was removed from the analysis, a non significant reduction in sway was still seen (t = 2.211, df = 5, p = 0.078) although a larger effect was observed (ES = 1.46)(Figure 6B). With all seven participants included in the statistical analysis, mediolateral sway was also not significantly different between the start and end of the trial (t = 2.270, df = 6, p = 0.064), with a corresponding large effect (ES = 1.55)(Figure 6A).

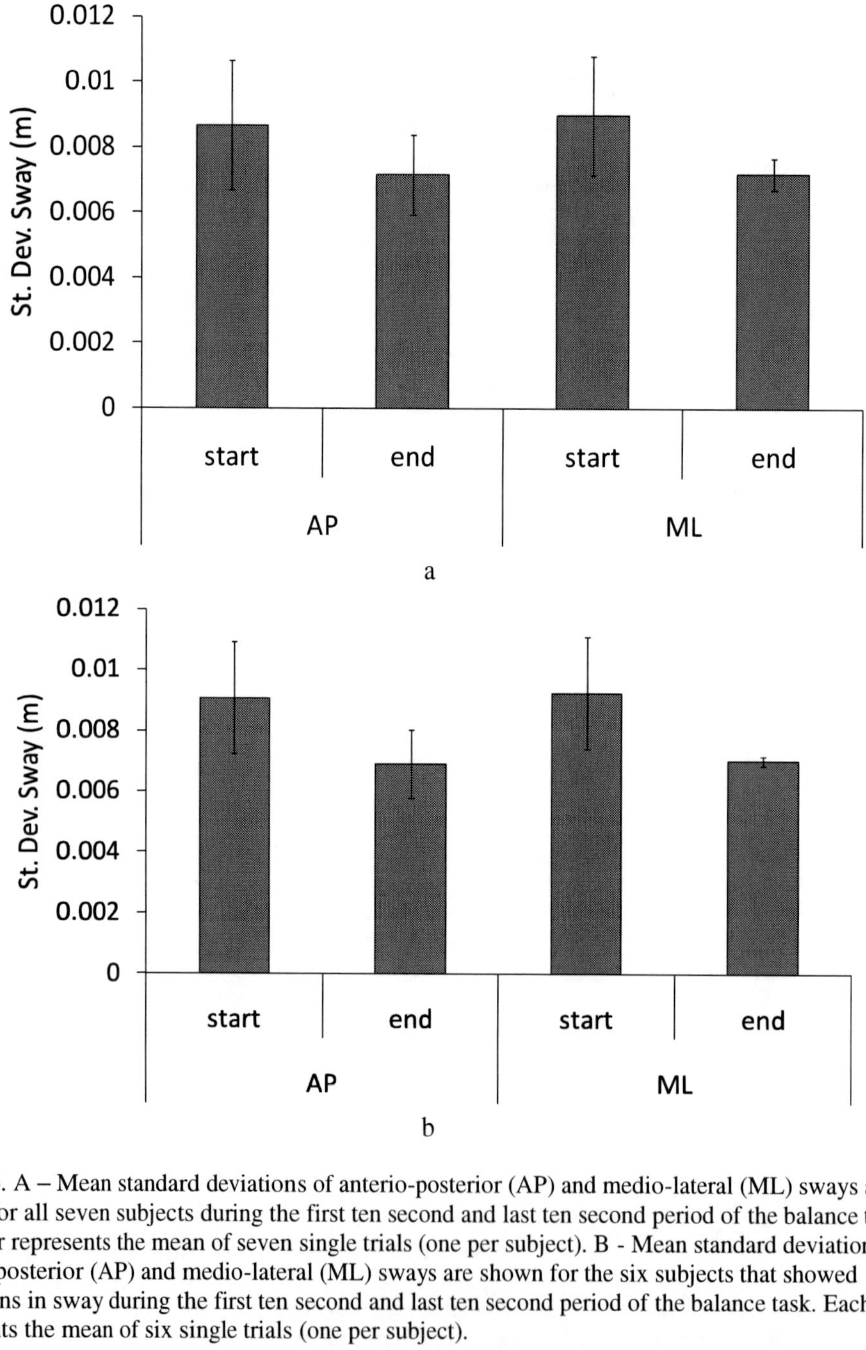

Figure 6. A – Mean standard deviations of anterio-posterior (AP) and medio-lateral (ML) sways are shown for all seven subjects during the first ten second and last ten second period of the balance task. Each bar represents the mean of seven single trials (one per subject). B - Mean standard deviations of anterio-posterior (AP) and medio-lateral (ML) sways are shown for the six subjects that showed reductions in sway during the first ten second and last ten second period of the balance task. Each bar represents the mean of six single trials (one per subject).

However, when the participant who showed an increase in ML sway was removed from the analysis, a more highly significant reduction in sway was observed ($t = 2.903$, $df = 5$, $p = 0.034$) with a very large effect ($ES = 2.25$)(Figure 6B). The subject who showed increases in AP and ML sway also showed the shortest duration for the maintenance of balance at 28.41 seconds, almost 20 seconds less than any other subject.

DISCUSSION

The aim of this study was to investigate the single leg sway characteristics of dancers whilst balancing on a rotating platform that was being used as part of a specific choreography they were training to perform. When the antero-posterior and medio-lateral sway characteristics were compared to each other, they were found to be similar during both the initial ($p = 0.766$) and terminal ($p = 0.938$) stages of the balance task. During normal (non rotational) single leg stance, medio-lateral sway is typically found to be smaller than sway in the antero-posterior direction (Hertel, et al., 2006; O'Connell, et al., 1998).

The similar levels of sway seen during the rotating balance task in the medio-lateral and antero-posterior directions could have a number of causes. Firstly, both the visual and vestibular systems will be perturbed when the body is rotated. The influence of rotation on the vestibular system will only last while the fluid in the semicircular cannals is accelerated. However, this perturbation of the visual system will be constant due to participants maintaining a constant positional relationship between the head and torso. Reductions in the amount of sway in both the antero-posterior and medio-lateral directions were seen over the course of the balance task, although these were only significant for ML sway. This reduction could support the suggested influence of the vestibular system in increasing sway upon the initiation of balance until steady state has been reached. Soeda et al. (2003) suggested that during rotation, any increased vestibular feedback in response to postural sway would detrimentally influence the participants' gaze stability. Thus, in the initial stage of the balance task performed here, the increased sway could be attributed to the already perturbed visual information becoming less useful as a result of vestibular perturbations effecting gaze stability.

Lackner and Dizio (1994) investigated the influence of whole body rotations in a fully enclosed room on postural control. It was found that after approximately one minute of rotation, the vestibular system had returned to resting levels and thus did not provide additional sensory cues. In the present task, the fluid in the vestibular system could have achieved "steady state" by the last ten seconds of the trial, potentially allowing gaze stability, and hence the processing of visual feedback, to improve.

Further support for the notion that the reductions in sway were due to a steady state being reached by the vestibular system is provided by St George et al. (2011). In their investigation, participants were rotated on a platform at an angular velocity of 0.34 $rad.s^{-1}$, which is very similar to angular velocity used in the present study of 0.38 $rad.s^{-1}$. Their participants gave an indication of perceived sway which increased at the start of the rotation and reduced to zero after 15.8 seconds. In the six participants who showed reductions in sway in the present study, the last ten seconds of the trial occurred after a minimum of 35 seconds. Thus, their perception of sway, influenced by the acceleration of the fluid in the semicircular canals (St George, et al., 2011), would have reduced between the start and end of the trial.

Another factor that would act to increase sway during rotation is the Coriolis force, C_F, which is given by

$$C_F = -2m(\omega \times v) \quad (3)$$

where m is the mass of the participant, ω is the angular velocity of the rotations and v is the linear velocity of the participant's centre of mass relative to the rotating frame of reference. In a series of studies investigating postural control in a fully enclosed rotating room, Lackner and DiZio (1994, 1998) showed that participants had reduced control of arm movements. Over time, the negative influence of the Coriolis force reduced as a result of adaptations of the sensorimotor system, and participants were able to perform planned muscle contractions in anticipation of the effects of the Coriolis forces. During whole body postural assessments, however, Soeda et al. (2003) did not see adaptations to the Coriolis force within four 25 second rotational periods. This makes it unlikely that such adaptations occurred in the present rotations which were of a similar total duration to those reported by Soeda et al (2003). If postural control adaptations did not occur, then it seems likely that the initial influence of vestibular feedback on gaze stability, as discussed above, was the primary cause of the increased sway during the initial ten seconds of the task. However, future investigations are needed to determine the specific role of each sensory feedback system on the control of stance during rotational standing tasks.

Previous research has commented on the dominance of the visual system in balance and how dancers show a reduced dependence on the visual system after training in comparison to non trained individuals (Golomer, et al., 1999; Hugel, et al., 1999; Perrin, Deviterne, Hugel, and Perrot, 2002). Hugel et al. (1999) showed that classical ballet dancers displayed increased effectiveness of the visual system in the maintenance of balance in comparison to controls. However, they found similar balance abilities when in the ballet specific "on pointe" posture in eyes open and eyes closed conditions. This suggests an increased effectiveness of the somatosensory system to compensate for a lack of visual feedback when the eyes are closed. During the rotational balance task performed in the present investigation, it could be that a reduction in the reliance on visual feedback would help to improve postural control during rotation. Indeed, Perrin et al. (2002) showed that in the sport of judo, where the athlete has to rely more on the somatosensory system for the maintenance of posture, balance control was found to be better than in dancers. If performing a rotational choreography, where the somatosensory feedback has to be used effectively, due to the constant perturbed state of the visual system, specific training of the somatosensory system would likely be beneficial.

Previously it has been shown that both ML and AP sway are increased through rotation during bipedal stance (Caplan, et al., 2009). This suggests an increased demand of rotational balance tasks on the sensory feedback systems, which could be useful for training and rehabilitation purposes. Rotational stance tasks could be incorporated into progressive balance training routines and this should be the focus of future studies. The role of the sensory feedback mechanisms will depend on the type of rotation performed. Although beyond the scope of the present study, it would be interesting to compare how balance performance differs between slow rotational tasks, such as used here, and faster movements such as the pirouette.

This study had a number of limitations. Due to the design of the rotating platforms that were being used by the dance company and were available for the study, it was not possible to directly measure the angular velocity of the platform. Therefore, it must be considered that the decomposition of the centre of pressure data into AP and ML sways was based on signal processing techniques. More accurate assessment of AP and ML sway could be achieved by integrating a force platform into the top surface of the rotating platform. Although our interpretation of the data suggests a role of vestibular feedback, in line with previous literature, it is not possible to say conclusively that this is the only mechanism influencing our findings. Future research should compare rotational stance to non rotating stance, as well as comparing eyes open and closed conditions. Finally, Only a small sample of dancers was available for the study. This could not be avoided as we required the dancers to be familiarised with the use of the rotating platforms for dance performance. It would be interesting to determine the effect of familiarisation to whole body rotations on sway, and this should be the focus of future research.

CONCLUSION

In conclusion, this pilot study has shown that the magnitude of medio-lateral sway is similar to antero-posterior sway when performing a single leg balance task when the body is caused to rotate slowly about a vertical axis. Reductions in postural sway were also seen over the course of the balance task, showing large effects, which were suggested as being attributed to vestibular adaptation. Further research is required on a larger population of dancers, however, in order to confirm the precise contributions of the postural control systems during this type of stance. The use of whole body rotation could be investigated as a training or rehabilitation intervention to stress the balance senses.

ACKNOWLEDGMENTS

The authors would like to acknowledge the professional dance company balletLORENT who provided the rotating platform used in the study. We would also like to thank Cinzia Hardy and Dr Sanna Nordin for their support throughout the design and implementation of the investigation.

REFERENCES

Caplan, N., Johnson, B., Horsburgh, S., and Walker, M. (2009). The role of sensory feedback mechanisms on sway size during rotating bipedal stance. *Gait and Posture, 30S*, S122-S123.

Cohen, J. (1988). *Statistical power analysis for the behavioral sciences* (2nd ed.). Hillsdale, N.J. ; London: L. Erlbaum Associates.

Golomer, E., Cremieux, J., Dupui, P., Isableu, B., and Ohlmann, T. (1999). Visual contribution to self-induced body sway frequencies and visual perception of male professional dancers. *Neuroscience Letters, 267*(3), 189-192.

Hertel, J., Olmsted-Kramer, L. C., and Challis, J. H. (2006). Time-to-boundary measures of postural control during single leg quiet standing. *Journal of Applied Biomechanics, 22*(1), 67-73.

Hugel, F., Cadopi, M., Kohler, F., and Perrin, P. (1999). Postural control of ballet dancers: A specific use of visual input for artistic purposes. *International Journal of Sports Medicine, 20*(2), 86-92.

Lackner, J. R., and Dizio, P. (1994). Rapid Adaptation to Coriolis-Force Perturbations of Arm Trajectory. *Journal of Neurophysiology, 72*(1), 299-313.

Lackner, J. R., and DiZio, P. (1998). Adaptation in a rotating artificial gravity environment. *Brain Research Reviews, 28*(1-2), 194-202.

Morrison, S., Hong, S. L., and Newell, K. M. (2007). Inverse relations in the patterns of muscle and center of pressure dynamics during standing still and movement postures. *Experimental Brain Research, 181*(2), 347-358.

O'Connell, M., George, K., and Stock, D. (1998). Postural sway and balance testing: a comparison of normal and anterior cruciate ligament deficient knees. *Gait and Posture, 8*(2), 136-142.

Perrin, P., Deviterne, D., Hugel, F., and Perrot, C. (2002). Judo, better than dance, develops sensorimotor adaptabilities involved in balance control. *Gait and Posture, 15*(2), 187-194.

Soeda, K., DiZio, P., and Lackner, J. R. (2003). Balance in a rotating artificial gravity environment. *Experimental Brain Research, 148*(2), 266-271.

St George, R. J., Day, B. L., and Fitzpatrick, R. C. (2011). Adaptation of vestibular signals for self-motion perception. *Journal of Physiology, 589.4*, 843-853.

Suponitsky, Y., Verbitsky, O., Peled, E., and Mizrahi, J. (2008). Effect of selective fatiguing of the shank muscles on single-leg-standing sway. *Journal of Electromyography and Kinesiology, 18*(4), 682-689.

Verhagen, E., Bobbert, M., Inklaar, M., van Kalken, M., van der Beek, A., Bouter, L., et al. (2005). The effect of a balance training programme on centre of pressure excursion in one-leg stance. *Clinical Biomechanics, 20*(10), 1094-1100.

In: Applications in Arts Biomechanics
Editors: Gongbing Shan and Peter Visentin
ISBN: 978-1-62808-015-5

Chapter 7

DIFFERENCES IN SKILLS THROUGH ANKLE JOINT KINEMATICS AND VERTICAL GROUND REACTION FORCE DURING DANCE MOVEMENT WITH POINTE SHOES

Yasuyuki Yoshida[1]*, Izumi Hamamoto[2] and Mayumi Kuno-Mizumura[2]

[1]The Center for Leadership Education and Research, Ochanomizu University, Bunkyo-Ku, Tokyo, Japan

[2]Department of Performing Arts, Ochanomizu University, Tokyo, Japan

ABSTRACT

The purpose of this study was to investigate the differences in the dance skills with pointe shoes by analyzing ankle joint kinematics and vertical ground reaction forces (vGRF) during the Rise-Up. A set of female subjects were divided into two groups—a pointe group of six experienced ballet dancers comfortable with using pointe shoes and a non-pointe group of eight dancers experienced in dance forms other than ballet. Two video cameras and one force plate were used to obtain the ankle joint kinematic data and vertical ground reaction force data. In each trial, each subject completed consecutive cycles of the Rise-Up. The results of the ankle joint kinematics demonstrated that the main difference between the two groups in their ankle joint movement was the duration of the total movement. In addition, the results of the vGRF provided further insights into the differences between both groups. A distinctively patterned rapid increase and decrease of vGRF values was observed in the non-pointe group during the upward movement. On the contrary, an irregular disorganized pattern was observed for the pointe group while performing the upward movement. During downward movement, numbers, timing and values of the peaks differ by group, an irregular disorganized pattern was observed for the pointe and the non-pointe groups. In the pointe group, the subjects did

* Corresponding and First Author's Institution: The Center for Leadership Education and Research, Ochanomizu University, 2-1-1 Bunkyo 2-214 Ohtsuka, Bunkyo-Ku, Tokyo 112-8610, Japan; E-mail: yoshida.yasuyuki@ocha.ac.jp.

not accelerate the center of mass rapidly, they acted the force to the floor in irregular output, so the movement of their center of mass were sensitively controlled while performing the both movements. The cause of characteristic peaks was rapid upward acceleration of subject's center of mass in the non-ponte group during the upward movement. We speculate that the bending stiffness of the shank might have caused the distinctive rapid increase and decrease pattern in the vGRF data for the non-pointe group. The dance skill required for the upward movement of the Rise-Up can significantly be improved for the non-pointe group by wearing pointe shoes. These findings may further benefit both practitioners and instructors of classical ballet and designers of pointe shoes.

Keywords: Dance skill, ground reaction force, kinematics, pointe shoes

INTRODUCTION

A pointe shoe is one of the most important tools for female classical ballet dancers. This shoe can help the dancer achieve the position of standing on the toe; this position is called a pointe position in classical ballet. Dancing in the pointe position is an essential skill for classical ballet dancers.

In previous studies, Weiss, Rist, and Grossman (2009) have suggested a set of guidelines for initiating pointe training. Further, Meck, Hess, and Helldobler (2004) have reported on the components involved in pre-pointe evaluation, i.e. evaluation of dance skills prior to and in preparation of pointe training. These studies indicate that classical ballet dancers should not wear pointe shoes in the initial stages of ballet training. In general, young classical ballet dancers begin to dance with pointe shoes after several years of ballet training. It is important to note that the acquisition of dance skills wearing pointe shoes is difficult for unskilled classical ballet dancers because classical ballet requires a dancer to pivot on the toes during the pointe position.

In classical ballet, one of the fundamental movements requiring the pointe position is called the 'Rise-Up'. The Rise-Up can be understood as a process involving three separate body movements or positions. The first movement in the Rise-Up is achieved by rotating only the ankle joints to create forces acting against the floor with the knee joints fully extended; this action is transformed into the vertical movement of the classical ballet dancer's centre of mass, thereby allowing the dancer to Rise-Up. When the ankle joints are fully plantarflexed, the classical ballet dancer is able to stand on the toes with the help of the pointe shoes; this is the second position. Subsequently, the ankle joints are dorsiflexed, allowing the dancer to return to the initial position. Thus, on the basis of ankle joint kinematics, the Rise-Up can be divided into the following three parts: upward movement, pointe position and downward movement. Although the Rise-Up is composed of fundamentally simple movements, it is not understood that what dance skill is required for the classical ballet dancers when they start to wear the pointe shoes during the Rise-Up.

It is noteworthy that the movements involved in the Rise-Up have not been investigated to date. In addition, biomechanical studies on dance movements performed while wearing pointe shoes are very limited. Lin, Su, and Wu (2005) investigated the symmetry between the dominant and non-dominant legs during the 'Releve', which is the Rise-Up performed with the flexion and extension of the knees. Chockley (2008) compared the ground reaction forces involved between the vertical jumps of the ballet dancers landing on the pointe position and the jumps landing on the full foot. In these previous studies, the subjects were sufficiently skilled in dance movements practiced and acquired by wearing pointe shoes. Therefore, these studies did not focus on the influence of pointe shoes in the performing of dance movements. Thus, the investigation of the biomechanics of dance skills practiced and acquired by wearing pointe shoes would shed light on the parts of the Rise-Up that are particularly difficult for unskilled classical ballet dancers.

The purpose of this study is to investigate differences in dance skills with pointe shoes through biomechanical analysis during the Rise-Up, thereby leading to potential improvements in the learning of particular dance movements.

METHODS

Subjects

In this study, fourteen female subjects were divided into two groups. One was a pointe group. This group comprised 6 subjects (n = 6) with an average age, height and mass of 20.83 ± 0.75 years, 1.61 ± 0.06 m and 49.57 ± 5.28 kg, respectively. The pointe group had 15.83 ± 2.40 years of experience in classical ballet including 11.83 ± 3.06 years in dancing with pointe shoes. These subjects wore their own preferred pointe shoes during the test sessions. The other group was a non-pointe group. This group comprising 8 subjects (n = 8) with an average age, height and mass of 20.63 ± 0.92 years, 1.60 ± 0.04 m and 50.65 ± 3.85 kg, respectively. The non-pointe group had 12.88 ± 4.39 years of experience in dancing styles such as modern dance, contemporary dance, tap dance and jazz dance. None of them had any prior experience wearing pointe shoes. This group wore identical sets of pointe shoes (Serenade, Bloch, Australia) during the test sessions. All subjects were required to sign an informed consent form approved by the institutional review board.

Instruments

Two digital video cameras (IXY M5, Canon, Japan) operating at a frequency of 60Hz were used to record the movements of the subjects and the control objects. A force plate operating at 1000Hz (9287B, Kistler, Switzerland) was used to obtain the vertical ground reaction force (vGRF) data. The cameras surrounding the force plate recorded movements in the right sagital and frontal planes of the subjects. In order to time-synchronize the cameras and the force plate, a rubber ball was manually dropped on the force plate in each trial.

Protocol

The origin and usage of certain classical ballet terms are region-dependent. In order to avoid confusion in usage of such terms, the generic term Rise-Up was used in this study. In each trial, each subject completed eight consecutive cycles of the Rise-Up with the both feet on one force plate. Each trial involved the subjects performing movements as described below.

First, the subjects assumed the first position in classical ballet technique, standing with the posteriors of the heels touching and each leg externally rotated at the hip to an angle of 60°. This was considered the initial position. Next, the ankle joints of the subjects were fully plantarflexed, allowing them to stand on their toes for duration of two seconds. Subsequently, the ankle joints were dorsiflexed, thereby allowing the subjects to return the initial position in an interval of two seconds. Each trial involved eight such upward and downward movements with the pointe position being assumed in between the movements. During the Rise-Up trials, the knee joints were kept fully extended. The right hand was placed on the hip, the left arm was extended to the wall and the left index finger touched the wall to maintain balance. A metronome operating at 60 bpm recorded the time intervals involved in the movements.

Data Processing

For each trial, the records of the third to sixth cycles of the Rise-Up were used for further analysis; consequently, four cycles of the Rise-Up were analyzed in this study. The records of the trials in each camera corresponding to the centre of the third metatarsalphalangeal (MP), ankle and knee joints on the right leg were manually digitized by using software (Frame-DIAS2, DKH, Japan). The records of the control objects were also digitized for direct linear transformation (DLT) methods (Abdel-Aziz and Karara, 1971) to obtain three-dimensional (3D) coordinate data.

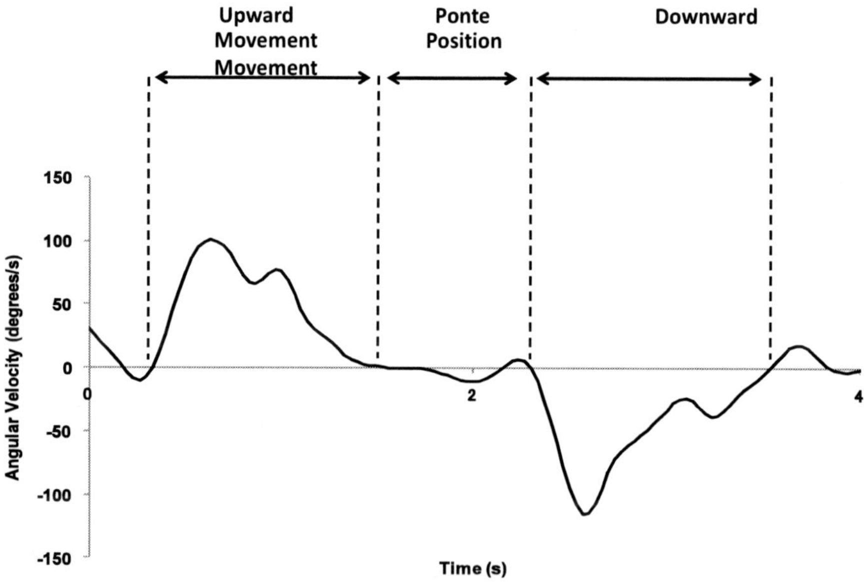

Figure 1. Definitions of upward movement, pointe position and downward movement during Rise-Up in one cycle determined by using angular velocity. Positive direction indicates plantarflexion.

Finally, the 3-dimensional position data of the above anatomical landmarks were reconstructed for all subjects. The digitized two-dimensional (2D) video coordinate data from the two cameras and the vGRF data were synchronized using the event of the contact of the rubber ball on the force plate. The coordinate data were smoothed by using a fourth order low-pass Butterworth digital filter (Winter, 2005) with a cut-off frequency between 2 Hz and 6 Hz. The vGRF data were normalized to the body weight of the subject.

The angle and angular velocity of the ankle joint were calculated from the 3D coordinate data. The angle for the ankle joint was derived from the dot product of the vectors from the ankle to the third MP joint and to the knee joint. The angular velocity for the ankle joint was obtained by differentiating the angle with respect to time.

Figure 1 provides the definition of the upward movement, the pointe position and the downward movement in terms of elapsed time during the Rise-Up in one cycle as determined by using the angular velocity of the ankle joint for kinematic analysis. The following is the detailed description of three parts. During the upward movement, the ankle joint was planterflexed, so the ankle joint experienced positive angular velocity. Subsequently, the angular velocity returned to zero and further assumed a negative value. Firstly, the maximum positive angular velocity was found. From that point, initially detected zero crossing points were searched in forward and inverse directions. This section consisted of these two zero crossing points were defined as the upward movement. During the downward movement, the ankle joint was dorsiflexed, so the ankle joint experienced negative angular velocity. Subsequently, the angular velocity returned to zero and further assumed positive values. The maximum negative angular velocity was found, and then the same approach as the upward movement had done. The pointe position was defined between the end of the upward movement and the beginning of the downward movement.

An independent t-test was carried out to assess differences in the peak values of the plantarflexion angle, the angular velocity for the ankle joint and the vGRF values between the pointe and non-pointe groups during the upward and downward movements of the Rise-Up. Furthermore, the durations of the upward movement, the pointe position, the downward movement and the total movement (calculated as the sum of the durations of the three parts) were assessed for both the groups. For each subject, the data over the four cycles of the Rise-Up were averaged. The statistical significance value was set at $p < 0.05$.

RESULTS

1. Ankle Joint Kinematics

Figure 2 shows the representative example of the angle and the angular velocity plotted against time for the ankle joint over the four selected cycles mentioned in the previous section. The two groups exhibited similar patterns of the ankle joint angle throughout the cycles. In addition, no significant differences were observed between the pointe and non-pointe groups in the peak plantarflexion angle values (Table 1).

During the upward movement, the ankle joint started to plantarflex from the starting position, and subsequently, it reached the fully planterflexed position (pointe position) for the pointe and non-pointe groups. There was no significant difference between both groups in the duration of the upward movement (Table 2). During the pointe position, the measured angle for the ankle joint was not completely stable; it varied slightly for both groups. Further, there was no significant difference between both groups in the duration of the pointe position (Table 2). During the downward movement, the ankle joint moved from the pointe position and dorsiflexed to return the starting position. There was no significant difference between both groups in the duration of the downward movement (Table 2).

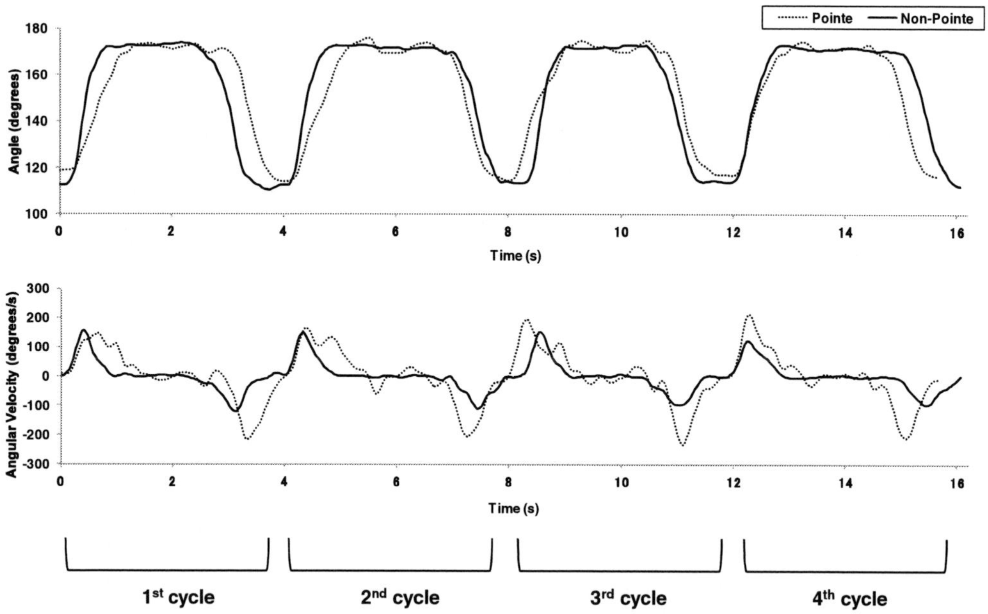

Figure 2. Representative example of the angle (top) and the angular velocity (bottom) plotted versus time for the ankle joint during the Rise-Up in four cycles. Positive direction indicates plantarflexion in the angular velocity.

Table 1. Mean (SD) peak values in the plantarflexion angle (degrees), the angular velocity (degrees/s) and the vGRF (N/BW) during the Rise-Up

	Pointe	Non-Pointe
Plantarflexion Angle	173.06 (2.69)	172.12 (4.65)
Upward Angular Velocity	132.47 (48.48)	170.58 (86.60)
Downward Angular Velocity	–121.33 (62.20)	–137.10 (76.21)
Upward vGRF	1.12 (0.04)*	1.22 (0.08)*
Downward vGRF	1.13 (0.02)	1.18 (0.08)

* indicates significant value difference between the groups ($p < 0.05$).

Thus, no significant differences were observed for both groups in the individual duration of all three movements. However, the duration of the total movement was greater for the pointe group than for the non-pointe group ($p < 0.05$). This is shown in Table 2.

There was no significant difference in the peak angular velocity values for the ankle joint in the upward and downward movements for both groups (Table 1).

Table 2. Mean (SD) values in the duration (s) of the upward movement, the pointe position, the downward movement and the total movement during the Rise-Up

	Pointe	Non-Pointe
Upward Movement	1.41 (0.22)	1.21 (0.35)
Pointe Position	0.66 (0.38)	0.54 (0.51)
Downward Movement	1.63 (0.29)	1.58 (0.30)
Total Movement	3.69 (0.19)*	3.33 (0.35)*

* indicates significant value difference between the groups ($p < 0.05$).

During the upward movement, the angular velocity patterns were similar for both groups until the first peak was reached. After the first peak, a few more peaks were observed in the case of the pointe group. In contrast, the angular velocity smoothly decreased for the non-pointe group. During the downward movement, the angular velocity patterns were the reverse of the upward movement.

2. Vertical Ground Reaction Force

Figure 3 shows the representative example of the vGRF data plotted against time obtained for the four cycles corresponding to Figure 2.

During the upward movement, the vGRF graph was different for the pointe and non-pointe groups. In the pointe group, an irregular disorganized pattern was observed while the subject performed the upward movement (plantarflexion).

This pattern had a few characteristic peaks. In contrast, a distinctive rapid increase and decrease pattern was observed in the non-pointe group. When the non-pointe group started the plantarfexion, the vGRF value increased and then decreased rapidly. As shown in Table 1, the peak vGRF value was greater for the non-pointe group than for the pointe group ($p < 0.05$). Figure 4 shows the plots of the peak vGRF values during the upward movement in all trials for both groups.

The peak vGRF value ranged from 1.03 N/BW to 1.22 N/BW in the pointe group and from 1.07 N/BW to 1.44 N/BW in the non-pointe group. During the pointe position, the vGRF values slightly oscillated about a central value for both groups. However, the average vGRF value remained almost unchanged for most of the duration of the pointe position. During the downward movement, the behavior of the vGRF patterns was similar for both groups. The irregular disorganized pattern, which was described in the upward movement of the pointe group, was observed for both groups during dorsiflexion. There was no significant difference in the peak vGRF value for both groups during the downward movement (Table 1).

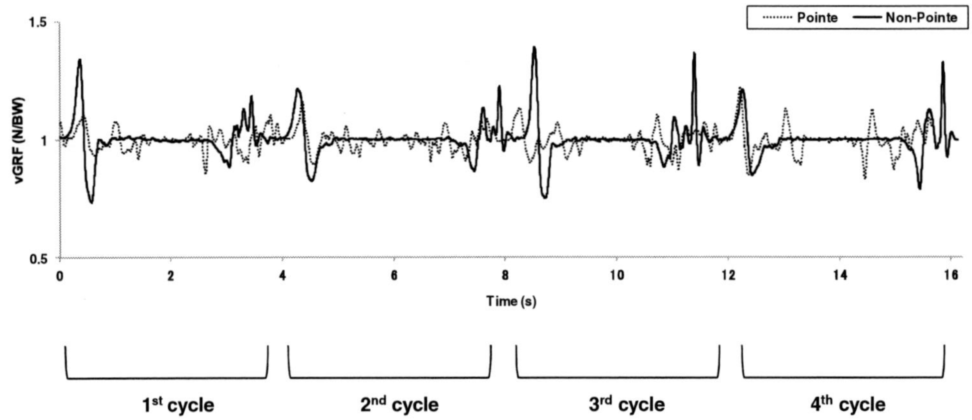

Figure 3. Representative example of the vGRF in the pointe group and the non-pointe group plotted versus time during the Rise-Up in four cycles.

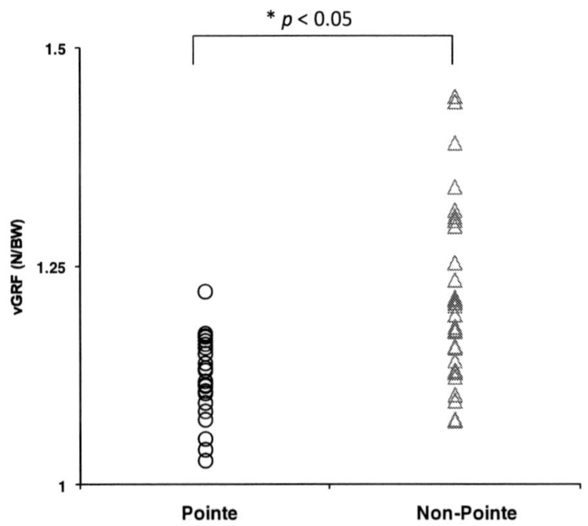

Figure 4. Plots of peak vertical ground force (vGRF) values during upward movement in all trials.

DISCUSSION

The analysis of the angle for the ankle joint showed that there was no statistical difference in the peak plantarflexion angle for both groups. This result suggested that the non-pointe group could achieve the pointe position while wearing the pointe shoes without having the dance skills practiced wearing the pointe shoes. The Rise-Up cannot be evaluated only by the dancer's ability to achieve the peak plantarflexion angle; the dancer's ability to smoothly execute the movement leading to the pointe position must also be considered.

The angular velocity patterns observed were not similar for both groups. Close to the end of the upward movement and the beginning of the downward movement, a few peaks were observed in the pattern for the pointe group. However, because manual digitizing was used to obtain coordinate data, small changes are not suitable for comparison. In this study, it is difficult to determine whether or not these small changes are related to the dance skills requiring pointe shoes.

The results of the angle and the angular velocity analysis for the ankle joint demonstrated that the main difference in the ankle joint movement pattern for both groups was the duration of the total movement. The duration of the movement was longer for the pointe group than for the non-pointe group, although the movement for all subjects was regulated by a metronome. For the pointe group, this longer movement may be the method used by classical ballet dancers for performing aesthetic movements while wearing pointe shoes. On the other hand, the non-pointe group had never previously worn pointe shoes; thus, they were not sufficiently skilled in the control of the ankle joint while wearing pointe shoes. Here, the definition of control is a very broad term. It includes not only factors of strength but also of coordination, timing and flexibility (Gans, 1985). In order to acquire the skills for dance movements performed with pointe shoes, the duration of the total movement should be changed for the non-pointe group. Thus, the non-pointe group needs to exert further control of the ankle joint in order to move longer during the Rise-Up.

The vGRF patterns obtained in the analysis of the Rise-Up constitute a very important factor. In this study, the vGRF patterns better revealed the differences in both groups when compared with the results of the analysis of the ankle joint kinematics. The analysis of the vGRF data showed that the differences between the pointe and non-pointe groups were present only in the upward movement and not in the downward movement. Characteristically, the pattern of distinctive rapid increase and decrease was observed only in the upward movement of the non-pointe group. During the upward movement in the non-pointe group, distinctive four peaks were found in 1.34N/BW, 1.21N/BW, 1.39N/BW and 1.21N/BW at 0.36s, 4.27s, 8.53s and 12.25s in Figure 3. respectively. The time-course of reaction forces, especially those of the ground reaction force resulting in acceleration of the system's centre of gravity, entails essential information about system dynamics (Ertelt and Blickhan 2009). The cause of these characteristic peaks was rapid upward acceleration of subject's center of mass in the non-ponte group.

On the contrary, an irregular disorganized pattern was observed for the pointe group while performing the upward movement. Considering the instruction for the subjects and the ankle joint kinematics, time range in plantarflexion corresponded approximately to from 0 to 2 s , 4 to 6 s, 8 to10 s and 12 to 14 s. in 1st cycle, 2nd cycle,3rd cycle and 4th cycle, respectively. The curve of the vGRF in these time was irregularly changed up and down for a brief instant of time. Since this curve is irregular, there is difficulty in viewing the peaks clearly as compared with the pattern of distinctive rapid increase and decrease. This difference is extremely definitive for the movement of center of mass in the both groups. In the pointe group, the subjects do not accelerate the center of mass rapidly. They act the force to the floor in irregular output, so the movement of their center of mass are sensitively controlled.

As reflect differences in the movement of the center of mass, the peak vGRF value in the upward movement was significantly greater for the non-pointe group than for the pointe group in the statistical analysis. Hugel, Cadopi, Kohler, and Perrin. (1999) reported postural control of classical ballet dancers during the pointe position; in particular, they reported that the Rise-Up is a particular skill that involves a specific technique. However, their research focused only on the static dance position and not on dynamic movement. In this study, we observed that dance skills practiced and acquired by wearing pointe shoes were also required during the execution of the upward movement.

During downward movement, numbers, timing and amplitude of the peaks differ by group, an irregular disorganized pattern was observed for the pointe and the non-pointe groups. Considering the instruction for the subjects and the ankle joint kinematics, the time range in dorsiflexion corresponded approximately to from 2 to 4 s, 6 to 8 s, 10 to12 s and 14 to 16 s. in 1st cycle, 2nd cycle, 3rd cycle and 4th cycle, respectively. As well as the upward movement in the pointe group, the curve of the vGRF was irregularly changed up and down for a brief instant of time while the both groups were performing dorsiflexion. The position of the center of mass has been going down to the floor as time advanced in the downward movement. Meanwhile, the center of mass has not been straightly falling down. They acted breaking force to the floor, irregularly. As a result, the vGRF curve became the irregular disorganized pattern.

Two different patterns were observed in the vGRF data. The vGRF curves were affected by the method of performing plantarflexion and dorsiflexion. Thus, the controls activating the planterflexor and dorsiflexor muscles significantly contribute to the vGRF pattern. In addition, we cannot ignore the method of transmission of the exerted forces (exerted by the plantarflexor and dorsiflexor muscles) to the floor. The forces were applied to the floor via the foot segment. It is to be noted from an anatomical point of view that the foot segment is a multi-joint structure and not a rigid structure. Typically, only two markers are placed to track the foot segment in biomechanical analysis. However, the information obtained can be misleading when motion occurs between the two markers (Leardini, Benedetti, Catani, Simoncini, and Giannini, 1999). Therefore, the vGRF curve may be determined by a complex interaction of many bones and muscles around ankle and foot during the Rise-Up.

Some previous studies have reported that classical ballet dancers are prone to overuse injuries occurring in the foot segment. According to Albisetti et al. (2010), localized stress fractures occurring at the base of the second or third MP joint are particular to classical ballet dancers. Moreover, Lo et al. (2007) reported an injury occurring at the distal phalanx of the big toe in a professional classical ballet dancer. Further, Nilsson, Leanderson, Wykman, and Strender (2001) have reported that classical ballet dancers working in the pointe position or the half pointe position are prone to injury caused by excessive load on the joints in the foot. These researches indicate that ballet dancers frequently move the joints in the foot segment while dancing with pointe shoes.

A distinctive rapid increase and decrease pattern was observed in the non-pointe group during the execution of the upward movement. Besides, the peak vGRF value in the upward movement was significantly greater for the non-pointe group than for the pointe group. In the upward movement the vGRF pattern was affected by the method of performing plantarflexion. At the beginning, the forces exerted by the plantarflexor muscles were transmitted to the foot segment. Then, the forces were transmitted to the shanks (soles) of the pointe shoes. After that the forces were finally transmitted to the floor. The shank of the pointe shoe, fabricated using cardboard, leather or their combination, helps in supporting the foot while in the pointe position by providing a certain degree of stiffness (Cunningham, DiStefano, Kirjanov, Levine, and Schon, 1998). This stiffness is called the bending stiffness (Oleson , Adler, and Goldsmith, 2005). Stefanyshyn and Fusco (2004) reported the bending stiffness of the sprint shoes appears to have an influence on sprint performance. We speculate that the bending stiffness of the shank might have caused the distinctive rapid increase and decrease pattern in the vGRF data for the non-pointe group. However, it may be difficult to prove whether or not the bending stiffness is exactly related to the vGRF in this study. One thing is for certain, the movement of the Rise-Up is complexly intertwined with related factors, such as the pointe shoe with the bending stiffness, the foot segment consisted of many bones and the forces excreted by plantarflexors and dorsiflexors.

The chief limitation of this study was the analysis of the foot segment. The simple rigid model was used as the foot segment because home-used video cameras were used for manual digitizing. This study also provided mainly descriptions in the vGRF and the joint kinematics, not provided sufficient for classify patterns. Differ from simple analysis by using standard parameters such as the maximum value and the duration, classification analysis for time-course pattern is highly complex. However, our study is the first to report the biomechanical analysis of the Rise-Up performed with pointe shoes.

CONCLUSION

To summarize, it was observed that dance skills practiced and acquired by wearing pointe shoes appeared to affect the execution of the Rise-Up for subjects of both the pointe and non-pointe groups wearing pointe shoes. This study showed differences in the movement patterns of the ankle joint kinematics and the exerted vGRF between the pointe and non-pointe groups. These results indicate that the dance skill required for the upward movement of the Rise-Up can significantly be improved for the non-pointe group by wearing pointe shoes. These findings may further benefit both practitioners and instructors of classical ballet and designers of pointe shoes.

ACKNOWLEDGMENTS

This study received financial support from KAKEN (22700614).

REFERENCES

Abdel-Aziz, Y.I., and Karara, H.M. (1971). Direct linear transformation from comparator coordinates into object space coordinates in close-range photogrammetry. ASP Symposium on Close Range Photogrammetry, *American Society of Photogrammetry*, Falls Church, VA.

Albisetti, W., Perugia, D., De Bartolomeo, O., Tagliabue, L., Camerucci, E., and Calori, G.M. (2010). Stress fractures of the base of the metatarsal bones in young trainee ballet dancers. *International Orthopaedics*, 34, 51–55.

Chockley, C. (2008). Ground reaction force comparison between jumps landing on the full foot and jumps landing en pointe in ballet dancers. *Journal of Dance Medicine and Science*, 12, 5–8.

Cunningham, B.W., DiStefano, A.F., Kirjanov, N.A., Levine, S.E., and Schon, L.C. (1998). A comparative mechanical analysis of the pointe shoe toe box. An in vitro study. *American Journal of Sports Medicine,* 26, 555–561.

Ertelt, T., and Blickhan, R. (2009). Describing force-patterns: a method for an analytic classification using the example of sledge jumps. *Journal of Biomechanics,* 42, 2616-2619.

Hugel, F., Cadopi, M., Kohler, F., and Perrin, P. (1999). Postural control of ballet dancers: a specific use of visual input for artistic purposes. *International Journal of Sports Medicine*, 20, 86–92.

Gans, A. (1985). The relationship of heel contact in ascent and descent from jumps to the incidence of shin splints in ballet dancers. *Physical Therapy*, 65, 1192–1196.

Leardini, A., Benedetti, M.G., Catani, F., Simoncini, L., and Giannini, S. (1999). An anatomically based protocol for the description of foot segment kinematics during gait. *Clinical Biomechanics,* 14, 528–536.

Lin, C.-F., Su, F.-C., and Wu, H.-W. (2005). Ankle biomechanics of ballet dancers in relevé en pointé dance. *Research in Sports Medicine,* 13, 23–35.

Lo, S.L., Zoga, A.C., Elias, I., Peterson, J.R., Chao, W., Green, J., and Morrison, W.B. (2007). Stress fracture of the distal phalanx of the great toe in a professional ballet dancer: a case report. *American Journal of Sports Medicine,* 35, 1564–1566.

Meck, C., Hess, R.A., and Helldobler, R. (2004). Pre-pointe evaluation components used by dance schools. *Journal of Dance Medicine and Science*, 8, 37–42.

Nilsson, C., Leanderson, J., Wykman, A., and Strender, L.-E. (2001). The injury panorama in a Swedish professional ballet company. *Knee Surgery, Sports Traumatology, Arthroscopy*, 9, 242–246.

Oleson, M., Adler, D., and Goldsmith, P. (2005). A comparison of forefoot stiffness in running and running shoe bending stiffness. *Journal of Biomechanics*, 38, 1886-1894.

Stefanyshyn, D., and Fusco, C. (2004). Increased shoe bending stiffness increases sprint performance. *Sports Biomechanics*, 3, 55-66.

Weiss, D.S., Rist, R.A., and Grossman, G. (2009). When can I start pointe work? Guidelines for initiating pointe training. *Journal of Dance Medicine and Science*, 13, 90–92.

Winter, D.A. (2005). Biomechanics and motor control of human movement (3rd ed.). New York: John Wiley and Sons.

III – Visual and Media Arts

In: Applications in Arts Biomechanics
Editors: Gongbing Shan and Peter Visentin
ISBN: 978-1-62808-015-5

Chapter 8

MUYBRIDGE, MOTION ANALYSIS AND ART

Xiang Zhang
Department of Physical Education, Xinzhou Teachers University, China

ABSTRACT

Human movement is an essential component in the representation and creation of art. As such, movement has been used in artists' conceptual models for hundreds of years. With the development of science and technology, more and more alluring aspects of human movement have been revealed by using motion analysis technology. Since Muybridge, such captured beauty has inspired artists' thinking to create numerous types of innovative art works.

Using examples, this article aims to demonstrate such influences on innovative thinking and the creation of art.

INTRODUCTION

Both artists and scientists have had an interest in human motion. Various human movements have been used as artists' resources and models for over 1000 years. It is not an overstatement that human experience, including movement, is a contingent condition of both the creation and cognition of art. For artists, depictions of human movement have had the particular challenge of representing time-based events in a single image (Shan, Visentin and Harnett, 2010).

Pre-photography, movement was represented by capturing a stop-action visual effect, for example Diskobolos of Myron (Figure 1). From motion analysis' point of view, the stop-action (i.e. the gesture) can hardly match the essence of throwing a discus in reality. But from an artistic point of view, its affects need not be accurate anthropokinetically. Art works may be merely representational, relying on iconography, cultural loading and for the experiential knowledge of the viewer (Shan, Visentin and Harnett, 2010).

Edward J. Muybridge (1830-1904) was the first scientist who systematically developed equipment and techniques to photograph the human movement, from bipedal gait to human athletic and dancing activities (Figure 2 and 3).

The time exposure and multi-frame stop action photographs of Muybridge during the post 19th century sparked controversy in the art world, as realist painters became conscious that their works were not so true-to-life, and Muybridge's work using human subjects and multi-camera stop action photo initiated a scientific approach to human movement representations.

Figure 1. *Diskobolos* of Myron (circa. 460 BC).

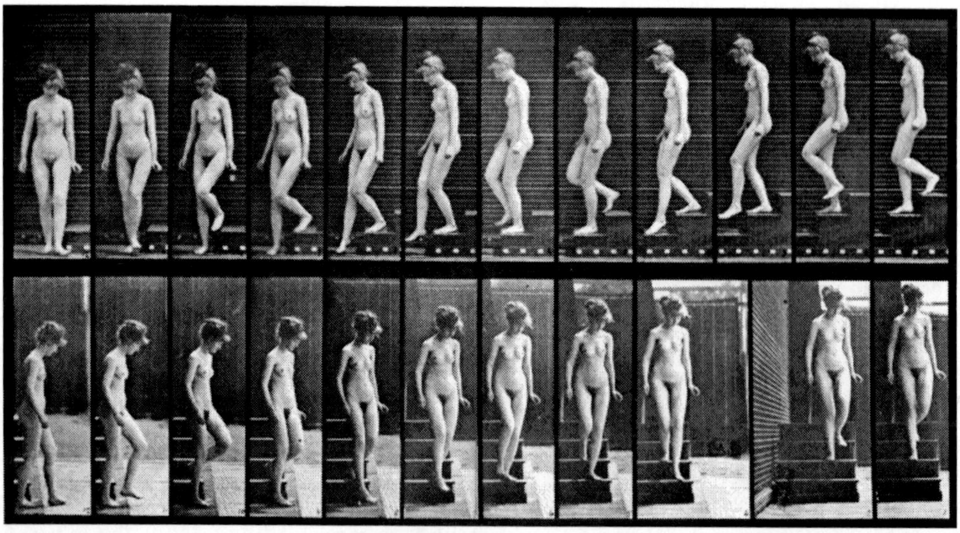

Figure 2. Example of Muybridge's work: Human bipedal gait (downstairs' movement).

Figure 3. Example of Muybridge's work: Jump.

The aim of this paper is to elaborate the influence of science on artistic creation, especially the influence of Muybridge and motion analysis technology. It'ss aim is to demonstrate that the advancements of science can help the union of both fields.

MOTION ANALYSIS AND ART

Muybridge's work using human subjects and multi-camera stop action photography initiated a scientific approach to human movement studies (Cavanagh, 1990; Solnit, 2003). The important scientific nature of Muybridge's work is obvious, even after 100 years, Muybridge's photographs are still of value in demonstrating scientific facts. A recent study done by Harvard University scientists (Bramble and Lieberman 2004) appeared in the Nature prompted a *Nature* cover picture (an artwork), which was based on Muybridge sequence photos (Figure 4).

Muybridge's work is often considered as means of qualitative evaluation of human movement rather than as merely as "art". However, the representation of age old concepts revealed in Muybridge' work provided new food for thought for artists. New ideas inspired new and to begin unique artistic creation. Such influence can been clearly identified in artworks detailing varieties of normal and abnormal gait, which look as though they were either sketched from his plates or photographed using methods similar to Muybridge's photograph technology.

One example is Nude Descending a Staircase painted by Marcel Duchamp in 1912 (Figure 5). One can clearly identify elements found in Figure 2 and 3 are applied in Figure 5. Essentially, such artists (20th-century Futurist painters) used techniques that compressed the principle of multiple exposure imagery into kaleidoscope art of force lines, vectored shapes and mobile patterns (Hunter, Jacobus and Wheeler, 2004).

Figure 4. The cover of *Nature,* November 2004.

Figure 5. Nude Descending a Staircase (Marcel Duchamp, 1912).

Figure 6. Dynamism of a Dancer (Gino Severini, 1912).

If traces of a "direct-copy" of Muybridge' photograph into a painting was more obvious in Nude Descending a Staircase, the artwork of Dynamism of a Dancer (Gino Severini, 1912) shows more independence (Figure 6). But, it still displays multiple fragmented essences in a single visual image – an idea derived from Muybridge. Nonetheless, this new approach of artistic representation - combining the traditional idea of stop-action affective representation with Muybridge's multiple exposure techniques - gives an expression of time and position designed to represent a perception of the human experience (Shan, Visentin and Harnett, 2010).

Many artworks are partially rooted in the representation of human movement. Such artworks can be viewed as a metaphor of the human experience. However, the advancements in technology have influenced both our existential perception of the world and artists' creation. Hence, the creation of art, influenced by Muybridge's photograph techniques, can be understood as a sequential expression of how we experience our current existence; one where conceptual extremes of this can be seen when the motion analysis technology effectively becomes "transparent".

Motion Analysis and Animation Art

Scientific analysis was at the forefront in Muybridge's. Despite the apparently dispassionate objectivity of Muybridge's images, clearly they have been the inspiration for representation of the emotions and passions of the human condition., The method, which examines human movement frame by frame, later was used by illustrators to draw sequences which when put together streamed as a moving body. Therefore, in addition to having an influence on painting, Muybridge's work also had great impact on film and animation.

In animation, there are two approaches for using motion analysis technology: 1) video tracing and 2) 3D motion capture. The first approach uses a technological devices (e.g. a rotoscope) that enables animators to trace live action movement frame by frame. Usually, animators assemble elements of a scene and insert characters whose movements are based on computer generated representations of real movement observed in video referents (e.g. tracings). . After tracing, the animated figured is coloured and then integrated with various background layers to create the final shot. Several Disney films employed such an approach to produce vivid scenery such as that found in Snow White and Peter Pan. (Figure 7).

The 3D motion capture, using mathematical filtering, has similarities to the process of video tracing. Both capture live human movement and use the captured data as referents for animation. But, 3D technology requires much less post-processing and manipulation to achieve the intended animation effects. The principle of 3D motion capture is to track movement of markers on human body. It does so by using multiple cameras to view and triangulate the positions of markers (normally passive reflective markers)typically placed on stable anatomical landmarks, such as joints. Using sophisticated software, precision motion capture measurements can be assembled in 3-D virtual space, supplying real movement condition for computer animation programs (Figure 8). The most advantageous characteristic of 3D technology is that it only captures markers which can be used to build the frame of a moving body (i.e. stick figure). Therefore, animators can create their figures first and then use captured 3D data to drive them. As such, they can "transfer" real human motion to their artworks (Figure 9). This type of practice supplies us with new aesthetic experiences.

Figure 7. Animation made via video tracing technology; Tinkbell peers through a keyhole.

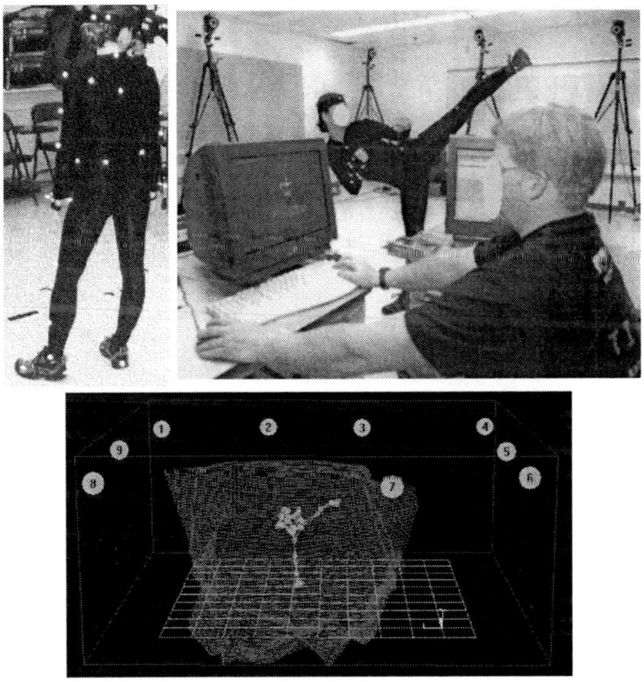

Figure 8. 3D motion capture: (top left) marker setting, (top right) data collection and (bottom) 3D reconstruction of the captured movement. Numbered circles are cameras used in 3D data collection and the blue curve lines shape the capture volume where 3D data can be accurately collected.

Motion capture is a useful tool for 3D animation. It combines art (figures made by artists) with science (the artistic figures held real performance through science and technology). It can be expected that the union of both fields will build more science-artists in the near future.

Figure 9. Artistic figures driven by human movements using 3D technology.

Motion Analysis and Visual Art

The time exposure and multi-frame stop action photographs of Muybridge initiated scientific representations of human movement. "If photography tested the liminality of visual culture in the early 20th century, 3D motion capture does so in ours" (Shan, Visentin and Harnett, 2010). One challenge is that cognition in these dynamic worlds relies upon the viewers to "… subliminally compress information from multiple images into a summative experience" (Shan, Visentin and Harnett, 2010). In this regard, 3D motion capture can help us to overcome our visual memory-limitations and it permits the accurate recording of human movement. The time-based movement trajectories of motion capture can be used as pattern-finding references for various human movements from sports to arts performance. A collection of such trajectories could used as a means to abstract the uniqueness of any human movement, just like a signature can be used to identify an individual.

> "Gesture communicates information, the meaning of which is driven by empirical experience. Perceptions of gesture as an unfolding of time-based events are intrinsically integrated with the form and function of human movement. These are keys to the aesthetic experience of interpreting gesture. When attempting to present the dynamic content of human gesture in a single still image, synthesis must occur; both time and bodiliness must be represented. These are the keys to reading the image. Since the viewer has experienced the phenomenon of gesture as a time-based event, interpretation should also be a time-based phenomenon. Informed by this, communication must be linked to the culture of the viewer, where iconography, visual analogy, and/or personal experience can play a role" (Shan, Visentin and Harnett, 2010).

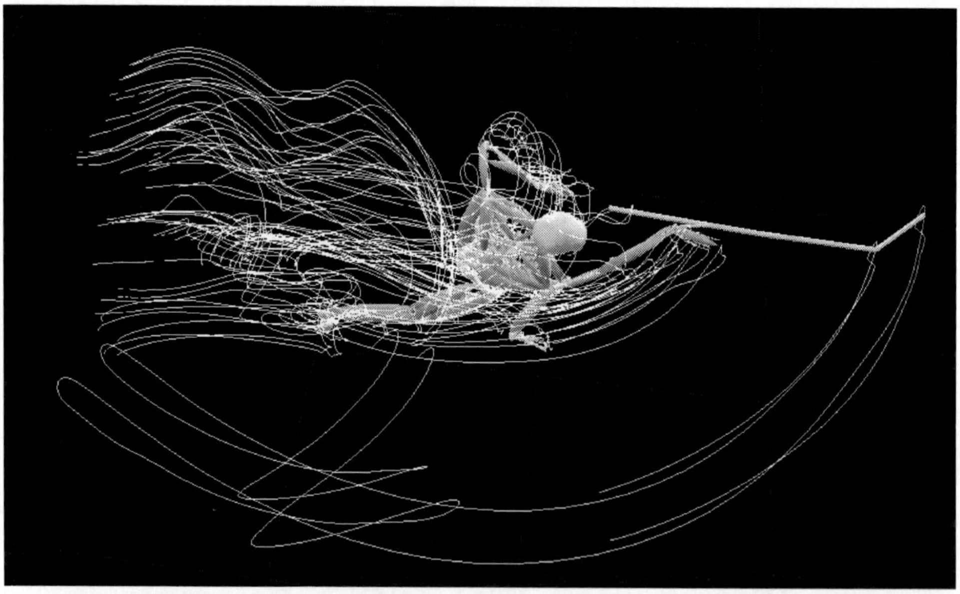

Figure 10. Signature of ice hockey shooting (top view).

Figure 11. Signature of baseball pitch (back view).

Figure 10 and 11 provide two examples of movement signatures generated using 3D technology. These exemplary prints should show the uniqueness of selected sport movements. They illustrate factors such as athleticism, power, and movement control. As such, motion capture trajectories can aid discourse about perceptual characteristics for us to explore and abstract the beauty of human movement. The objective nature of the movement signatures provide significant food for thought with regard understanding and analyzing human gesture as a means of communication, culture exchange and aesthetics exploration.

CONCLUSION

The influence of science on artistic creation, especially the influence of Muybridge and motion analysis technology on the visual arts, film and animation, is significant. Historical evidence has shown that scientific means can provide artists seeds for their artistic creation. It is could be concluded that, with further development of motion analysis technology and a creative ethic that values the intersection of science and the arts, more creative art works can be generated to enrich our culture life and increase human potential.

ACKNOWLEDGMENT

This work was supported by Shanxi Scholarship Council of China and Science Research Grant of Province Shanxi/China.

REFERENCES

Bramble, D. and Lieberman D. (2004) Endurance running and the evolution of *Homo*. *Nature*, 432, 345–352.

Cavanagh, P. (1990). The mechanics of distance running: a historical perspective, in P. Cavanagh (Ed.) Biomechanics of Distance Running. Champaign: Human Kinetics, P. 135-164.

Hunter, S., Jacobus, J. and Wheeler, D. (2004). Modern Art, Third Edition Revised and Expanded. New York: Prentice-Hall.

Shan, G.B., Visentin, P., and Harnett, T. (2010). A Novel Use of 3-D Motion Capture: Creating Conceptual Links between Technology and Representation of Human Gesture in the Visual Arts. *Leonardo Journal,* 43(1): 34-42.

Solnit, R. (2003). River of Shadows: Eadward Muybridge and the Technological Wild West. New York: Viking Penguin.

In: Applications in Arts Biomechanics
Editors: Gongbing Shan and Peter Visentin
ISBN: 978-1-62808-015-5

Chapter 9

MOUVEMENT CONCRÈTE; MODIFICATION OF BIOMECHANICAL MOTION DATA FOR NON-NARRATIVE AESTHETIC TIME-BASED EXPRESSION

William Smith
Department of New Media,
University of Lethbridge, Germany

ABSTRACT

This article explores how traditional biomechanical datasets created through motion, or performance capture technology can be modulated for aesthetic time-based means/ends. At the same time since human motion normally is abstracted from a bipedal figure, and that figure is pictured at a place and time, secondarily-seeming considerations are addressed related to the presentation of motion featured as 3d animation digital choreography.

Keywords: Aesthetic biomechanics, art biomechanics, dance technology, non-commercial mocap application, temporal manipulation of motion, 3D animation variation, temporal tweaking of mocap keyframes, digital choreography

I. INTRODUCTION

Time-based Art Forms

Movement composition requiring electronic technology developed in the last two decades shares commonalities with other forms of time-based expression - such as film editing, starting in the 1890s; electronic music editing starting in the 1930s; and video editing starting in the 1970s. One of these, *musique concrète* from the currently designated area of electronic music (a musical movement developed in the '30s and '40s when sounds could be recorded and then used as elements for expressive exploration which was subsequently codified, inspiring many composers, in the '50s and later) provides a good model to understand developments of contemporary motion capture.

In one aspect of *musique concrète,* sounds from the same or different sources recorded on audiotape could be manipulated by being spliced next to one another. Additionally they could be slowed down, speeded up, reversed, and blended by using other devices. According to Schrader [1], Schaeffer, a pioneer in this technique, felt that "sounds could be removed from their usual context and changed by manipulation." Schaeffer employed five basic tape manipulation techniques: loops, cutting and splicing, speed change, direction change, and tape delay.

In hindsight, one may see loops as merely one application of cutting and splicing that forms a closed system, and is a compositional device related to rhythm or a relative time unit. It is not a method, per se. Nevertheless, all of these previous codified precepts for *musique concrète* have contemporary equivalents related to using motion capture (Mocap) data. The physical act of cutting and splicing has now been replaced in the digital realm in all time-based artforms facilitated by computers by 'cutting and pasting'. But, in fact, more transformation can occur than merely repositioning, as 'cutting and pasting' might suggest.

Mocap Method

At its present state, motion capture technology can determine the x,y,z of important joints (or, muscles in facial capture) of body, or rotational aspects (pitch, yaw, and roll of body parts), or both. Samples of spatial location are taken/recorded in time at speeds of up to 1200 frames per second (Hz), or more. After the datasets have been processed (so quickly that it may appear to be instantaneous - or, in other words, in 'realtime'), often through a rigorous process of reconstruction and clean-up, they can be observed from any vantage point close or far from a point on a sphere surrounding the figure displaying the data.

I use the term *dataset* as a collection of numbers at one sample (Hz) related to sensed objects on the body. If I am using an optical system that calculates the x, y, and z (3 degrees of freedom) of each marker, and my subject (talent) is wearing 53 markers that are tracked, one *dataset* then contains 53x3 or 109 numbers. (It actually contains more, as a *dataset* has its own sequence number.) The plural of *dataset* is *datasets.*

I also use the term realtime to suggest a nearly instantaneous connection with the action of the biological source - the person doing the moving - and the representation of that source doing movement in the digital realm. The representation can be merely a stick-like figure, or can have more surface, seeming corporeal. It can even be a representation of a figure conceptualized for the final rendering of an animation process. Of course, there is computer processing going on, so there might be a lag of microseconds or milliseconds. Thus realtime is subjective; what is realtime for one system and group of people may be different for another. It is about perception. If there is a slight lag in time, an event in the digital space might 'seem' still as happening at the moment.

Thumbnail Mocap History

Motion curiosity is expressed in the last quarter of the 1800s in the work of Muybridge, and this connects with the imagination of those currently investigating motion. He used banks of cameras to record actions for sequential analysis and then compared individual frames to each other. For a summary, see Shan et al [2]. Rotoscoping used moving-picture film to record action and after developing and enlarging the images, the individual frames were guides to hand-drawn (cell) animation. Such a technique is said to have guided parts of the animated Disney feature, Snow White (1937). Images over time were also analyzed during the '80s by kinetic research teams using dots on joints and measuring the white spots on video image frames relative to other key points and then plotting them on graphs. Technologies driven by the need to visualize movement in 3d made their appearances in the '90s, and presented a host of solutions - including magnetic, electromechanical, optical, and others.

Mocap Use Today

Mocap is widely used in scientific research. It is used to maximize sport performance, for biomechanical analysis, for perceptual research, to engineer ergonomically designs, and in the medical profession.

A much more widely known application of mocap is in forms of entertainment [3]. Games use precaptured motion datasets to drive digital characters in interactive play, but also use realtime motion capture of the player/user in some systems. Performance Animation in realtime has been a regular process of TV aimed at young audiences since the early '90s. Realtime mocap is also found in some VR immersion. Mocap datasets have also been mapped on digital characters in feature films such as *Titantic*, *Lord of the Rings*, and *Avatar*.

Sampled mocap datasets mapped to 3-D digital models is often part of the story-telling process, but some people explore motion in non-commercial and non-functional applications that are outside of normal/typical interests. Exploring aesthetic movement can be seen as doing something for the sake of its own, and not using action to reach a goal or accomplish some task. For instance, the muscular actions of plucking an apple are undertaken to achieve the goal of having an apple. This is not normally considered aesthetic movement. The action one undertakes in this example is not about the motion, but of the result of motion, something other than motion itself.

Moreover, the motion examined here is not in service of telling a narrative story. It is not a means to reinforce with body motion signifiers such as speech or emotional states. In other words, it is not a referent. It is a primary element of human expression, that is, of digital movement composition.

Preproduction, Production, and Postproduction

Before I share methods of temporal tweaking (using techniques to alter time of an objectively captured sequence of motion datasets) or other techniques I have been using for about a decade, the reader should have some awareness of contextual considerations.

Conceptual planning initiates any execution; research starts with an idea. It can be framed as a question, for instance, "What happens when I explore a particular biomechanical movement sequence?" For instance, I was interested in resisting gravity through all forms of jumps, but mainly through sissonnes (starting on two feet and landing on one) as the movement basis of the artwork entitled *3 Marys, RGB*. In other works, I was (pre)interested in figure formations, color, and movement speed, as in "section 4" of *Journey*. In still others I explored gestural shapes, creating a movement palindrome combining non-realistic keyframed interjections with mocap datasets, and having a dynamic environment that featured colored objects (mostly primitive boxes) that changed position or size (e.g. in *Purity* the boxes were used as counterpoint to the white humanoid figure mapped with the mocap data).

There are two scenarios I have faced when assembling a digital motion work - I either already have the mocap datasets I will work with, or I must create them in order to achieve my digital movement goals. The latter suggests a preconception of what I might have at the end. I confess I don't always know what to expect when I start working with mocap datasets, and . . . such are the results of experimentation that cross science and the fine arts - and some succeed. Of course there is little utility in publishing a failed experiment, and I would wonder why anyone would ever be interested in seeing one.

Some work in this area may start out with a predetermined figure, or environment, and then think about motion. Others can start with motion. There can be no rule in creative processes. Frankly, it might not be important, but in the end, the parts that create the whole need to be orchestrated. They need to work together to bring about the responses in those experiencing the work that were intended by the creator(s). It is possible and likely that a certain combination of parts could evoke imagery/connection to those experiencing an animation that could be more powerful than another combination of parts, or a variant.

There is a synergism of original intent, production experimentation and post-production editing. One tries to get it right prior to starting production. If one is engaged in a process and trying to get it right, one may shift methods to seek a best practice. Finally there is review, reflection, and revision if necessary. At the end, one is trying to make sure the presentation of the movement connects (as I find similarly is going on with the materials/content I am thinking about here, and you, the reader).

Decisions for 3D Aesthetic Expression/Perception with Mocap Movement

Movement: Sometimes a movement question may dominate the research space. That movement data can be pre-existing or can be created within established research parameters. I have done both, with *Broken Egg* - using mocap data I didn't create, and for instance, with *Gatineau* - where I was really interested in visual polyphony expressed sonically through my research in deconstructing visual imagery into numerical data. I created three sequences of mocap datasets each dependent on a rhythm.

Figure: Any human movement is likely to be represented by a human-like figure with weight supports on two legs. Human movement could be mapped on a tyrannosaurus rex (although I am not certain that I would ever have any interest in doing this). Human perception is also likely to take an interest in the protagonist regarding gender. In some of my work I have chosen male (*Look Where You Land*), female (*Arcadian Women*), and non-sexual/gender (*Journey* and *Getting Business Done*).

And there can also be a sense of age. In *Nuclear Fancy,* a work that is earliest family portrait in movement I am aware of, I asked my daughters and wife to choose the models to represent them. It is interesting that my youngest chose a model older than her and my eldest (at that time, six) choose a baby model and performed movements consistent with that model. My wife, whom I met while she was performing in *Nutcracker* (her movements trained through repetition are balletic) had been retired for about 20 years at the time of the mocap session, so her actions reflected this reality at the moment of the capture session. In short, this work is sensitive to the fact that mocap datasets represent a unique body with unique experiences with unique mental and other capacities at a certain moment of that body's existence. And, of course, that same body can provide variants in movement over time.

Figure 1. "Gatineau" From Drumming Ottawa from Space(s) Beyond (2007).

As well as male or female, neutral (generic humanoid) figures dominate my work *Blue____onTheBlackSpan*. But they morph - from an animated contorted fabric surface, humanoid figures emerge, then turn into abstracted butterflies. Figures don't have to remain constant.

Figure 2. "GettingDoneTheBusinessOfWest,Central,North,AndEast" From Drumming Ottawa from Space(s) Beyond (2005).

Figure 3. Nuclear Fancy (2003).

Now while representation aspects may be literal or figurative, some of the early aesthetic explorations of mocap movement consisted of painted lines on invisible bodies in *Hand-drawn Spaces* and *Biped* (both by Cunningham), or of glowing dots or squiggles in Bill T. Jones' *Ghostcatcher*. This avoided conceptual issues of reductionism (going from more points to less) in the symbolic space of 3d animation represented early on screens often as 72 ppi and or resolution of 640x480.

While I am interested in two-legged digital models, others may choose non-bipeds. And, still others may map only parts of the mocap datasets, for instance, only the upper body data.

Number of figures: Sometimes, a statement cannot be made by one figure. *Ritual at the Well* seems communal. It was conceived/made at the time my father was dying of pancreatic cancer that blocked his stomach from connecting to his colon, and I was present as he starved to death, and could do nothing. There was a well attended wake, and, in the big scheme of things, I am aware that you, me . . . we all experience death, and there are emotions about loss. It seems so big. It seems (or maybe this is my projection/perception) that death of an individual affects many. It affects a community.

Yet, while *Armed with Reach* is (I think) a proactive statement for females, I don't think additional figures are necessary. One figure expresses the idea. I have daughters, and I want them to earn/seize what they need to be/feel successful.

Costumes for digital figures? Being in the arts, I've never felt uncomfortable with nudity. Nudes are found as statues or paintings going back thousands of years. Now, digital human models always are reductionist in that they have fewer points than a real person. I have never felt uncomfortable with nude digital models. I never felt that I had to clothe digital models simply because we wear clothes in our society.

I have mapped textures as in the *Virilian Hero Dance* that look like body suits that might be found on strange superheros found in comics or in made-for-the-populace wrestling or other contact sports.

Figure 4. Virilian Hero Dance (2002).

In *3 Marys, RGB*, I made the bodies of the figures invisible and only the rubberish-appearing body suits performed the mocap movement.

Environment: I will try to isolate this parameter, but I am not certain how to establish a hierarchy. Certainly environment is not the figure moving (read, foreground . . . often). It must be everything other than that figure represented as moving (and in my interests, that figure moving with mocap data that has been thought about and assembled, forming a composition with a beginning, middle, and end).

Figure 5. Section 8 of Journey (2006).

There is a tone . . . is it dark or light? Is it warm (reds) or cool (blues)?

Space: There can be a sense of limited space or unlimited space; often this is expressed through scenes inhabiting what we perceive as indoors or outdoors. *Broken Egg* was confined (feels that way when I see it, since the 3d space is so shallow/limited) and section 8 of *Journey* is also confined. At the end of section 8, the camera pulls back and the viewer can see the scene, like a theatrical set. Shakespeare who coined “All the world is a stage . . .” would no doubt have appreciated the layering.

Journey parts 10 and 13 are outdoors. The sky and sun signal that in part 10. The rain and growth of plants do so in part 13.

Figure 6. Section 10 of Journey (2006).

Time: Is it day or night? Certainly *Journey* part 6 was night (and outdoors, and big - there were many figures moving while sculptural letters were moving).

Figure 7. Section 6 of Journey (2006).

Journey part 2 is daytime with its brightness and shadows.

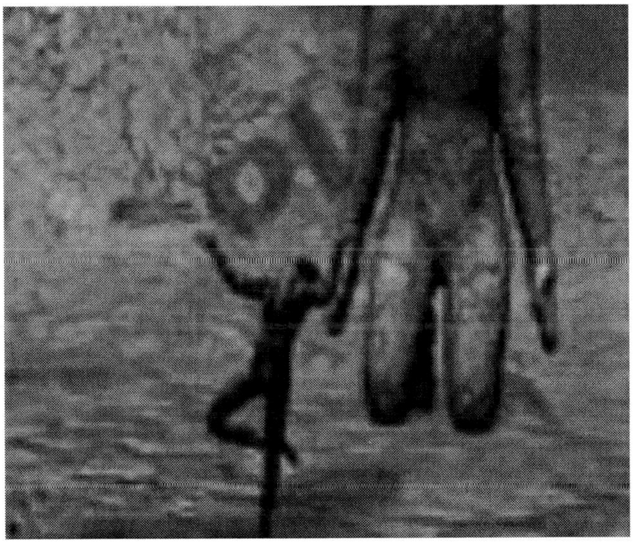

Figure 8. Section 2 of Journey (2006).

Objects can be active or passive. In *Look Where You Land*, the figure is jumping and objects like land sharks (actually they are cones suggesting impalement, signaling danger) congregate, menacingly; whereas in *shadeGreen*, the figure has a sword that is passive (moves with the hand).

Figure 9. Look Where You Land (2002).

In *Purity* utilizing shapes and colors that might be found in a period of Mondrian's artistic output in late 1919 and very early 1920s, environmental objects seem to interplay with the performer.

Moving-image projection/art. One can project movies within the digital space. I have never opted in my animations to execute this possibility, however I did this in a real embodied performance *Ten Year of Yours in the Eye of a Sleeping Dragon* (my statement about the process to get tenure).

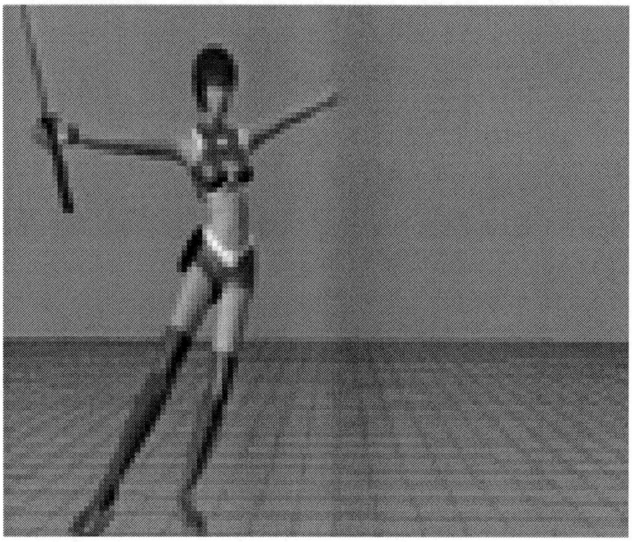

Figure 10. ShadeGreen (2002).

To me the concept is about time and how we measure it. The projection was of a time-worn animal trail through woods during an early spring/late winter snow. I was aware the deep past of nature and of cycles of weather that go beyond any human's experience. My experience in the workplace seemed inconsequential, pitiful compared to making meaning of life. The projection made sense and augmented the movement composition.

Figure 11. Ten Years of Yours in the Eye of a Sleeping Dragon (2002).

Special Effects

So many possibilities exist; if you can imagine it, well, you can probably create it digitally. It is assumed in this discussion that the special effects connect with the mocap movement and its contribution to artistic intent. In *skinlessNights*, fog existed in the one layer where a figure moved while a second one jumped overhead, landing on small platforms with a series of sparks. At one section of the work, trajectory trails are made from the hands, glowing at the x,y,z point of the 3d animation space lasting seconds before fading. It is about the imperfection of the concept of love, where we, as biological creatures, have DNA encoded genetic imprints on how we respond, yet the levels on which our intelligence, emotions, imagination, and education (maybe, awareness) are different as we partner for what we believe is a life-long journey. Well, . . . it is part of our human condition. So for me, the special effects (I had at the time) used for the work are germane for the visuals.

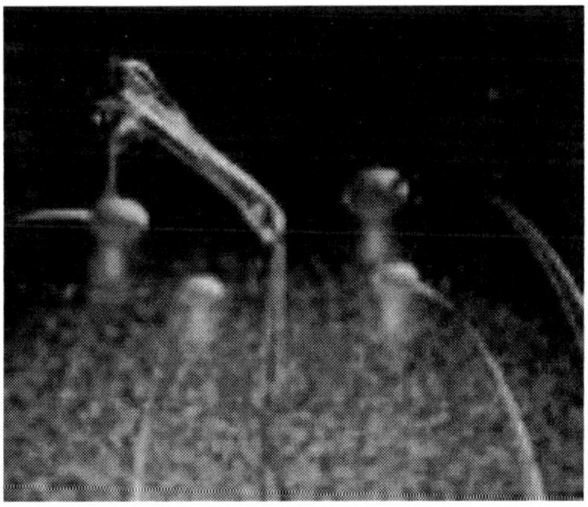

Figure 12. SkinlessNights (2003).

Animated Environment: I felt that putting a wave disturbance on the ground was appropriate to what I wanted to express in *UndertheSunOntheLand.* I wanted to world to be dynamic according to my understanding of how first nations people conceived of the world and of their place in it.

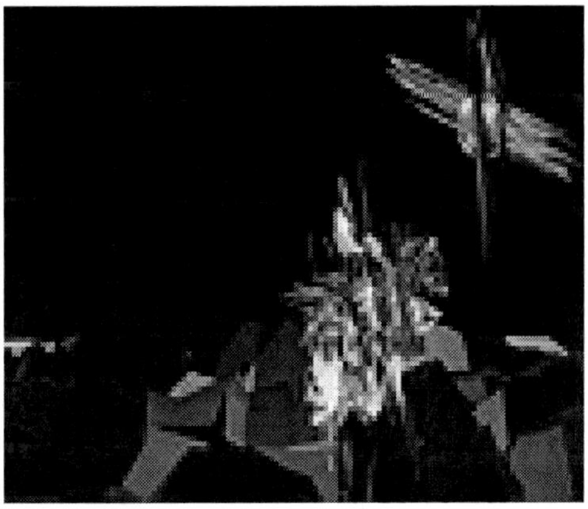

Figure 13. UndertheSunOntheLand (2002).

Camera: Limitless viewpoints; Close up seems powerful, far away seems reflective; can pan and dolly; can change cameras or have what I call the omnipotent camera that always sees the action as in *ThreeBlueRocks*. [4].

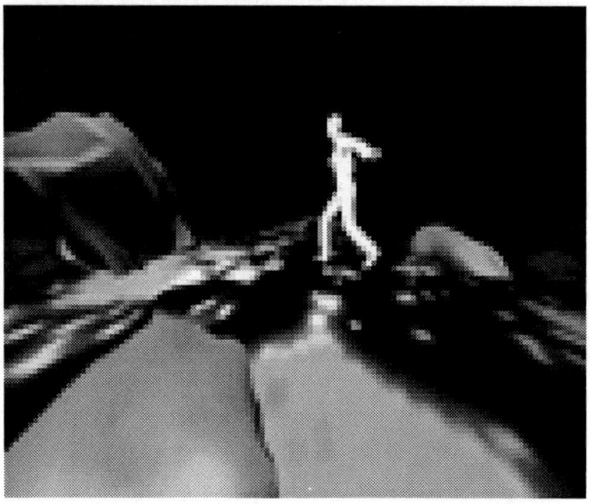

Figure 14. ThreeBlueRocks (2003).

The camera can be stationary or move, it can be fast or slow when it moves. If cameras change, pacing as in the length of shots is likely to be considered to maximize the expression. In Weiner's *Dialogos*, the camera moves quite a bit.

Sound: The discussion of making this choice to accompany mocap datasets far exceeds the scope of this article; but sound is very important. The same 3-D rendered movement sequence put to different soundscapes can produce strikingly different responses. To assert this does not mitigate the importance of figural representation or any of the other aspects for which a creator must make decisions to create the whole - that final mocap animation best expressing a movement-related intent.

Post-processing techniques: One can apply effects, filters, and a host of pixel manipulation techniques to a rendered frame, sequence, or entire animation. I did this to *storyTelller* and the final version of *Glasshouse*.

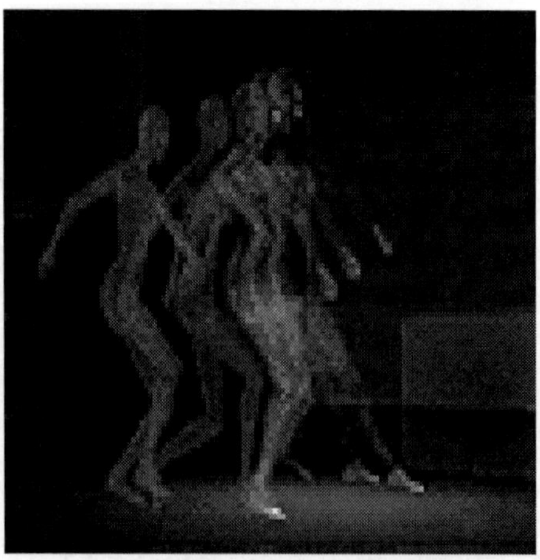

Figure 14. StoryTelller (2002).

I also heavily ghosted, desaturated parts, and tweaked the pixels in the xy grid in my keyframed *My Hand*.

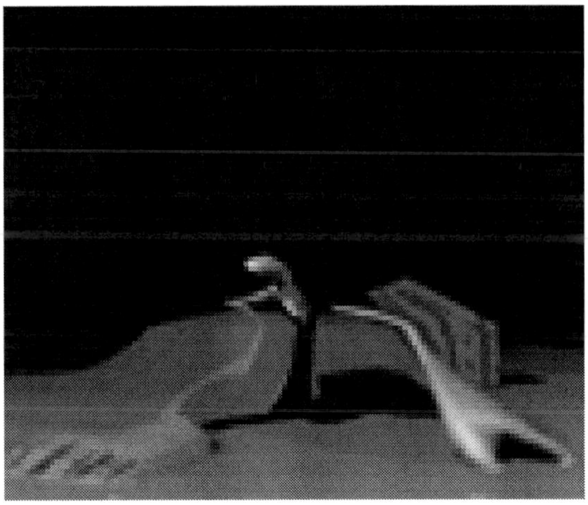

Figure 15. MyHand (1998).

All of these parameters and their possibilities are entwined with the movement choices that are discussed next.

II. MANIPULATING MOCAP DATA

As part of a methodology to tweak temporality of mocap datasets, I will discuss compressing, amplifying, progressing, reversing, snipping, blending horizontal, blending vertical, and repurposing, and other aspects. While capture is often at 120 Hz or greater, I am going to discuss the mapping of data captured at 30 Hz to the normal animation frame-rate for video at 30 frames per second. This I believe will make the theory/methodology more understandable (and it will reduce size and unnecessary complication in my diagrams).

Here is an illustration of this paradigm:

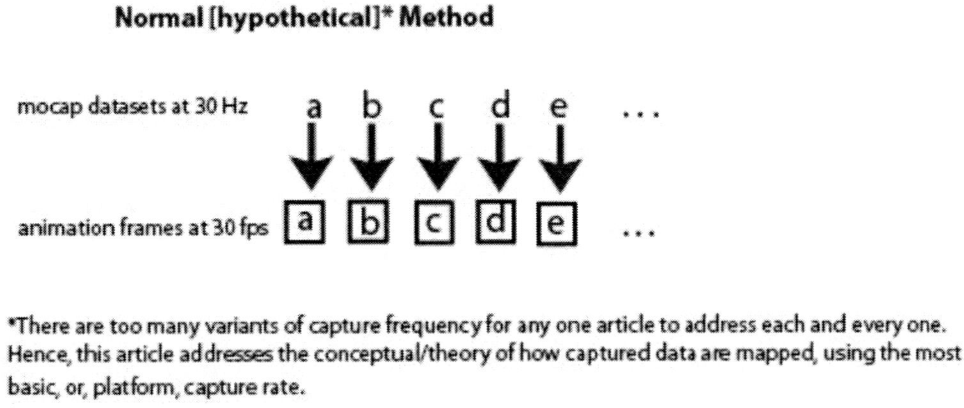

Figure 16. Normal (hypothetical) Method.

Compression (Faster)

This technique entails discarding mocap datasets. Consider a hypothetical sequence of 30 datasets where you keep the first dataset, discard the second, keep the third, discard the fourth, and so on. The result will be where you once had 30 datasets, you now have 15. Where the normal sequence took a second, the new sequence takes a half a second to get to the same body position. The effect to a viewer of an animation using mocap datasets in this manner is that the motion, originally normal to what a human could do, appears faster.

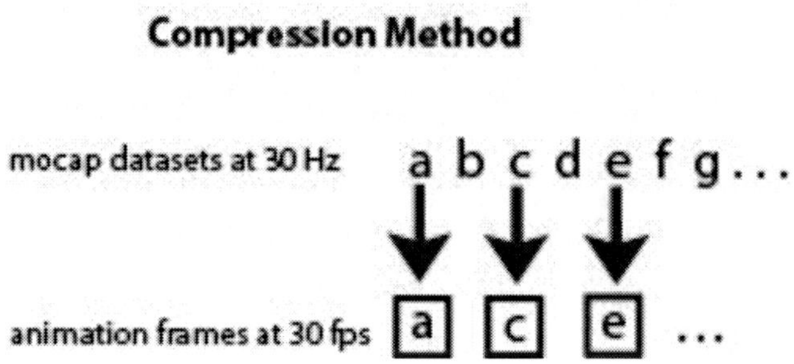

Figure 17. Compression Method.

Were we talking about a motion data captured at 120 Hz, we would keep the first datumset and discard the next seven. You can see why this discussion attempts to link a mapped motion-capture set at 30 Hz to normal 3d animation (and not feature films of 24 fps, rather as 30 fps which is a video standard.)

The technique above can be expressed as a mathematical mapping - where you have one set of numbers (normally designated by x) that is transformed by an equation (or mapped) to another set (normally designated by y). So if the sequence numbers of the mocap datasets mapped to animation frames is equated to y, then we could create a mathematical equation to express how to translate from mocap temporality x to animation temporality y: here, $y=2x-1$, creating ordered pairs of (1,1), (2, 3), . . ., that is for the first animation frame we want the 1st dataset; for the second animation frame, we want the 3rd dataset, and so on. And indeed, there is a way to express an algorithm for each of the following transformation methods of mocap data to animation timeline (amplification, etc), but mathematical expressions are not necessary to express concepts to readers who practice the artform, so this sample should suffice.

How many datasets you eliminate at a time will determine how much faster the result appears. For instance if you keep the first dataset, discard the second and third datasets, keep the fourth, discard the fifth and the sixth, you will end up with 10 where you once had 30. The new sequence takes a third of a second, and will appear faster than one that takes a half of second, and, of course, faster than one that appears normal (30 sets captured at 30 Hz played back as a 30 fps animation).

So you see how it is where you could create a new sequence of 6 datasets by maintaining a pattern of keeping one dataset and discarding the next four, or a new sequence of 5 by keeping one dataset and discarding the next five.

These above examples show constant incrementation within the mocap data, or, in other words, a good mapping since the sum of the kept and discarded datasets as a single pattern are factors of 30 (e.g. 2, 3, 5, 6) and one can see that 10 and 15 will also work well in the example of this article. But this is relative to an arbitrary temporal marker of 30 that divides a second of time. One should not be limited to thinking that a second is the only referent to establish a standard, for instance - as a measure line in a bar of music might establish a temporal unit of sound. In the end, one can discard any integer and end up with minor discontinuities, as is shown below. For instance, one can also discard three after keeping one datumset, and, at the end of 60 datasets, the cycle of the pattern will be completed, producing a new sequence of 15 datasets. The new sequence will take one fourth the time of its corresponding original. Any pattern can be established.

This technique of compression can be applied to all datasets, as seen in Weiner's *Dialogos* where mocap movement is accelerated by a factor of 2 [5].

Figure 18. Dialogos 2002.

Compression can be applied to only sections of a movement piece, not the entire work, and I also have played with one or more figures moving quickly at the same time one or more figures displayed normal biomechanical movement, as in *Virilian Hero Dance*, *Glasshouse*, and "section 4" of *Journey*. [6]

Figure 19. Section 4 of Journey.

I telescoped (really speeded up, about 3 minutes of movement animation mapped to 10 seconds) in *While You Were Sleeping*.

Figure 20. WhileYouWereSleeping (2003).

This was done in post production, discarding rendered frames, rather than in production, discarding datasets.

In summary for this method called compression, a viewer can perceive that certain motions have been speeded up. This is accomplished by a digital movement manipulator/choreographer using mocap data by reducing the datasets. In contrast, the following discussion of amplification of keyframes corresponds to this methodology by creating the condition of a perception of a slowing down of motion.

Amplification (Slower)

Using a methodology similar to making mocap movement appear faster by reducing data, one can make mocap movement appear slower by adding data.

So if you have a 30 Hz capture and you want make the time of the original to double, you merely duplicate each dataset. So where you had 30 frames in the animation, you now have 60 frames. Where you had a movement/action taking one second, now it takes two. This gives the appearance of the motion being slowed down.

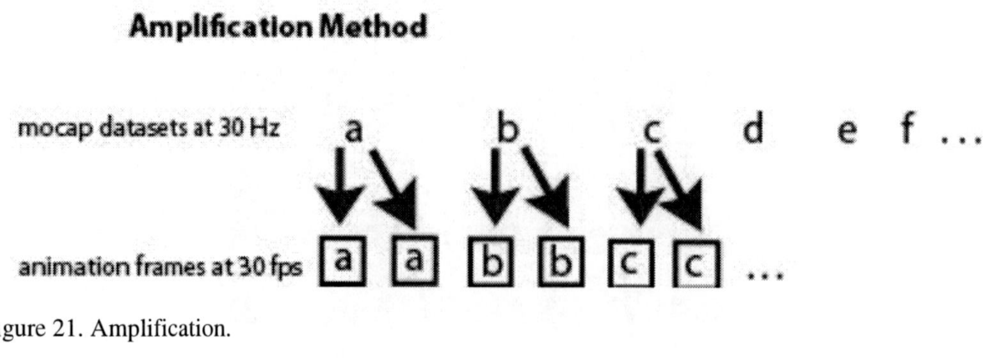

Figure 21. Amplification.

And were one to have each dataset appear three times in the animation timeline, then where the motion took one second it now takes three.

So one can double, triple, quadruple etc a dataset to create movement sequences that are outside of the norm, but serve the aesthetics of the animation expression.

If one has sufficient datasets, then one has to solve a mapping problem, not one of duplication of a dataset, as in the hypothetical 30 Hz captureset presented above in this discussion. For instance, in the major part of *Ritual at the Well*, I mapped a 120 Hz sequence of mocap data to 30 fps animation, essentially slowing it down to a fourth of its original speed. In other words, I mapped each dataset of mocap data to an animation frame, using 120 samples per second for 120 animation frames. [7].

Figure 22. Ritual at the Well (2007).

Progression

In general, a human-created biomechanical action has a normalacy to it that we can perceive by our eyes experienced in the laws of physics - that is - the way mass behaves in gravity. For instance, for a forward jump to occur there needs to be a change of angles of joints (bending of knees) forming the preparation, then some muscles contract and others extend, sending a person into the air for a period of time we can calculate with knowledge of weight and measurable forces at normal gravity (it is different in outer space), and then the body lands with some muscles and body parts absorbing the force like a spring. There is nothing magical about this, it happens according to laws of gravity. These laws supercede any desire for something else to happen - at least they do with embodied movement. We can defy natural/normal laws with mocap datasets in 3-D animations that are mutated.

Were there a wish to suspend the body in space, merely slowing it down at a particular moment to a sequence different from the rate of the first might appear abrupt; hence, one might want to treat the addition (or subtraction) of datasets in a non-linear manner. Such a technique gives the impression of deceleration (or acceleration.)

One can make such a sequence gradually slow down by a formula for a treatment of a datumset as in the following computer routine/logic for a sample of 10 datasets

x=0

Repeat to 10

x=x+1

end repeat

This would give a sequence of 1,2,3 . . . 10 that could be schematically shown in the diagram, producing a linear form of deceleration. The first dataset would get mapped to one animation frame, the second would get mapped to two frames, the third to three frames, and so on.

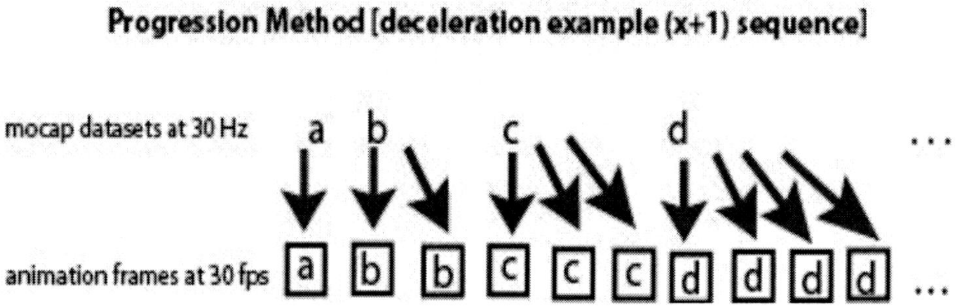

Figure 23. Progression Method x+1 sequencing.

One can find other schemes that might be more satisfying, such as seen below.

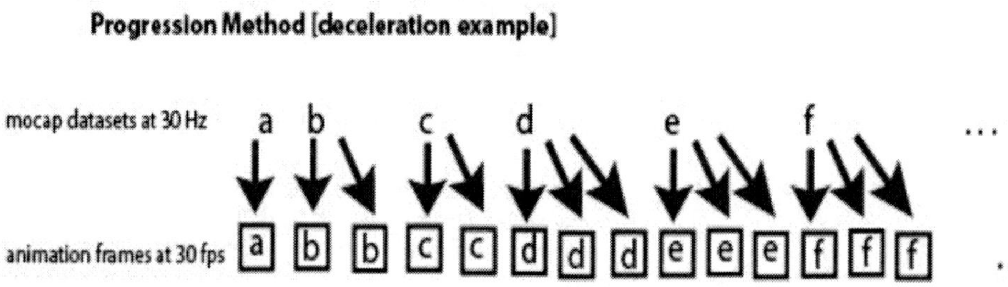

Figure 24. A Variant Schematic of the Progression Method.

A scheme similar to this was used in *3 Marys, RGB* to elongate the sense of suspension in the air. Now, while the initial assertion that mocap treatment could be envisioned similar to sound has largely been left unexplored to this moment, a normal sound has an attack, a sustain, and a decay. Comparing what I have done here with mocap datasets to *musique concrète* is to keep the attack, modify the sustain, and eliminate the decay. [8]

A discussion of phenomenological aspects as presented in Kozel [9] is beyond the scope of this presentation, but as someone who moved, was trained in some movement styles, and still moves, and has both memory and imagination of movement, it should be no secret that I felt the sense of eluding gravity at the time of creating the mocap datasets, and still do everytime I view the performance animation/digital choreography.

Figure 25. 3 Marys, RGB (2003).

Keyframing Mocap Datasets

Above, I've discussed the process of duplicating keyframes, that is, preserving the original dataset of the motion captured, however, one can map two successive datasets and produce other data, or nuances that didn't exist in the original motioncapture. To do this an animator spaces two successive datasets out so tweening occurs. (This corresponds in some sense to music's concept of microtonality.) How tweening or interpolation works in most 3-D computer animation programs is that a linear line is imagined between the same joint at two positions in the animation timeline corresponding to two (often successive) mocap datasets, and the line is divided into equal segments according to (the number plus 1) of frames added. So in the following example where three animation frames are created, a joint moves four times: a-ab1, ab1-ab2, ab2-ab3, and ab3-b. Once the equation of the line is determined (as part of the mathematics underlying the tweening process) it is then easy to identify the x,y,z coordinates of the joint, creating new data, for respective positions between the mocap datasets used as keyframes.

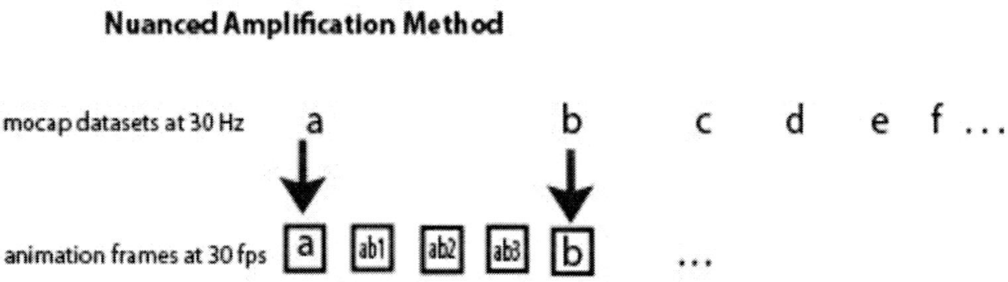

Figure 26. Using mocap datasets to generate new data.

Reversal

One can take datasets and change their order, putting the first one last, and the second-second to last, and so on.

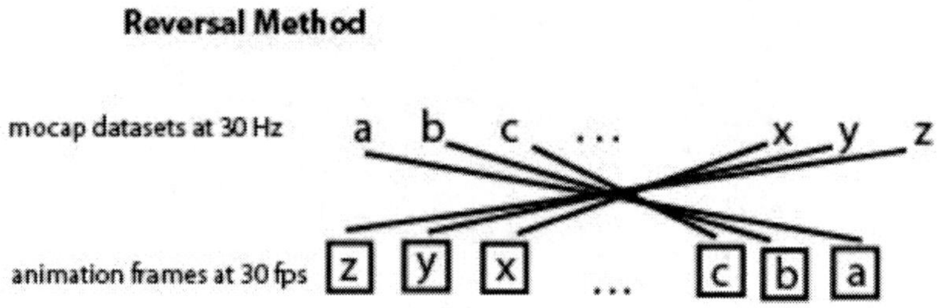

Figure 27. Reversal Method.

This can be for a substantial length of datasets as in *Purity* where for the first half (about three minutes) the 3d figure uses mocap datasets going forward, and in the second half, uses the same ones but in reverse order, creating almost a palindrome (it has some non-mocap keyframing in the first and second halves that are unique, so it cannot be considered a true palindrome).

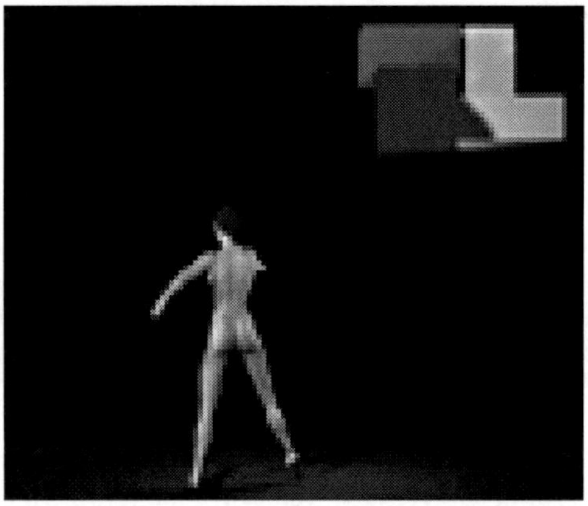

Figure 28. Purity (2002).

But this can happen also to shorter sequences of datasets, for instance, of twenty seconds as in *Virilian Hero Dance*. Such a method can be applied to a single movement idea as in *Glasshouse* and *storyTelller*. [10].

Figure 29. GlassHouse (2002).

Snipping to Discontinuity

When one wants to have an abrupt disruption of the laws of physics, one can put one mocap dataset sequence after another, without blending. This disruption can be in terms of movement, spatial location, or both as seen in *3 Marys, RGB* where the motion sequence at the height of the jump is cut off from the normal landing and the figure seems to magically appear in another location ready to begin another jump.

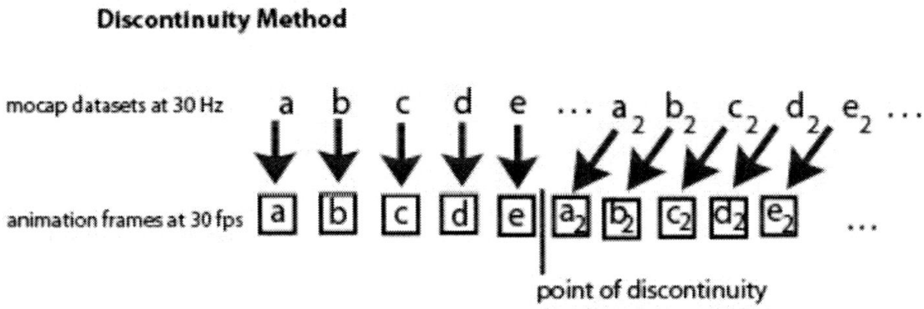

Figure 30. Discontinous Datasets.

Juxtaposing Non-mocap figure manipulation with Mocap datasets through keyframing

Sometimes it is necessary to use non-mocap data as part of a movement idea/expression. While a mocap dataset is actually a keyframe, keyframing is a term traditionally used in 3d animation discussion where a pose is established by the animator at one point in time, another pose is established later, and the computer determines through human programming, using mathematics, the spatial increments based on distance divided by frames (time) for corresponding body parts to move in their x, y, z dimensions. The computer-generated figure positions between two keyframes are referred to as the inbetween, or tweening.

So if one were to wish to insert 'passable as natural' or even physically impossible movement into a realistic mocap sequence, one could use the last dataset of a mocap sequence as one keyframe, set the beginning keyframe of the non-mocap sequence at a point later in time so that the computer could form an artificial, but smooth transition over several frames (suggested 10-40 usually), and use a similar process at the end of the inserted segment to return to mocap datasets.

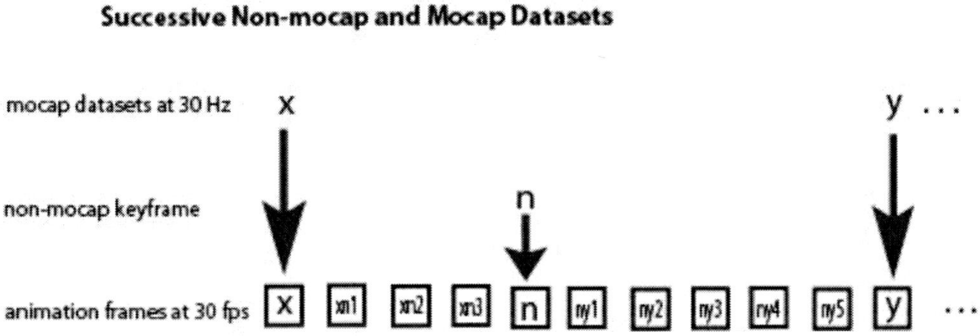

Figure 31. Sequencing non-mocap data with mocap data through keyframing.

I used this technique in *UP, Slammed Against the _______* which expresses my emotions/reaction to an administrator's abuse of power.

Figure 32. UP, Slammed Against the ________ (2003).

But I had also used it quite a bit in *Purity* [12].

Using Mocap Data for Non Traditional Figures

While one sees mocap data mapped on variants of a single figure, in *Blue____onTheBlackSpan*, I used one skin with a mocap sequence in canon (a canon has one time-based participant with a delayed beginning, such as heard in the song/round *Row, Row, Row your Boat*).

Figure 33. Blue_____onTheBlackSpan (2006).

The motion is not mapped directly to one recognizable figure; in other words, a figure might represent several mocap datasets at a single moment but not blend them to create a singular motion experience.

Repurposing mocap datasets created in one context for aesthetic ends.

Prior to having access to equipment where I could create my own mocap datasets, I was thinking about movement in 3-D spaces, and challenged myself to see if I could take existing data make choices, and make that data aesthetically interesting. This technique has connections to art movements of using 'found' objects as the basis for artistic expression. If one entertains the liberal construct that anything can be an element in an aesthetic construction, certainly motion can be too. I had access to many sport-related datasets, and used some to create *Broken Egg*.

In one section a figure climbs a wall from out of the ground while other action is going on in the foreground. Here you see the lower figure snowboarding [13].

Figure 34. Broken Egg (2002).

Any mocap sequence of datasets can be the substance for an aesthetic exploration/experiment/expression/composition.

I challenge all readers to think of a repository of mocap data, a free mocap library (maybe named after the Alexandrian library from classical times), where you can check out datasets and formulate new connections. And aesthetic movement need not codified as dance.

Local Gravity Referents of Moving Figures Non-Aligned with the World Gravity Referent

Another technique of using mocap datasets is to apply a pitch, yaw, or roll to figures contrasting that of the world, that is, based on the origin (what is the referent for what is up and down). While this can be simulated for an entire scene by rotating a camera, datasets can be out of sync with the axis of gravity that is portrayed in the 3d world. Who can forget those scenes in *Royal Wedding* (1951) that portrayed Fred Astaire dancing on walls and the ceiling, accomplished by rotating a room with the camera fixed as if setting on the floor, and Astaire appeared always keeping vertical as the box rotated. This can also be simulated in post-production where green-screened images can be keyed in scenes.

I explored the concept of modifying mocap data through rotation in *Non-Plain Planes* [14].

Figure 35. Non-Plain Planes (2002).

Having the mocap motion on figures simultaneously occupying the same space in the 3d world.

In our normal experience of the world, two objects cannot occupy the same space because of physical laws. An embodied movement producer cannot have joints in the x,y,z space of another embodied movement producer. Yet, in the 3-D space of an animation, boundaries are an illusion. This aspect is one I explored in *Arcadian Women* by using a canon of five figures starting in the same space at different times, and overlapping. Each figure has mocap datasets mapped to it, but delayed, and not delayed in a mindless manner. The delay is not in regular intervals. Moreover, each of the five figures associated with the mocap datasets has a certain degree of transparency, which is another parameter that adds interest.

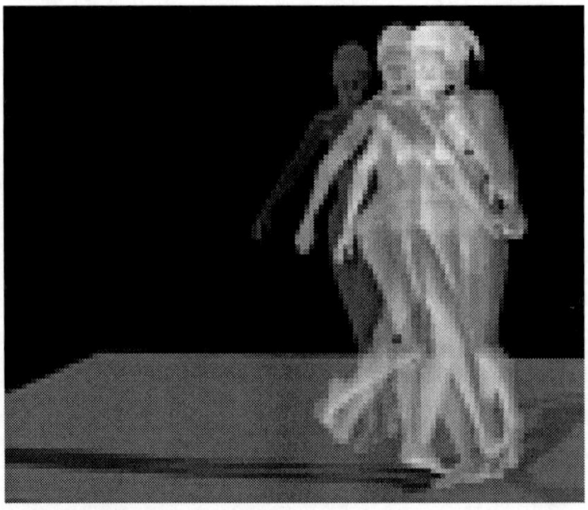

Figure 36. Arcadian Women (2002).

Also, in "section 12" of *Journey*, there are three groups in canon with each other, each doing a canon - and as a result of the mocap datasets where the body does not move much in the x-y plane, the figures appear to overlap. They seem to occupy the same space simultaneously. [15]

Figure 37. Section 12 of Journey.

This happens too at various moments in *Glasshouse* and *Storytelller*. In *Ritual at the Well* based on movement ideas constrained by the same parameters to create a unique, or a personalized embodied motion sequence by each of about 10 performers, the digital figures pass through one another. This could lead to a discussion later about 3-D as a symbol or an imitation of our normal experience. I don't think using mocap data in an aesthetic modality need ever to be constrained to laws of physics. I think we, as humans, have the imagination to be able to process movement/motion intent when it goes beyond the reality of normalacy that might be signaled by biomechanical reality. So while we can modify time, we can also modify space. And in the spirit of this discussion, certainly we can recognize patterns.

Linking/Blending Mocap Datasets in a Horizontal Manner

To create a plausible or acceptable movement sequence without borrowing terminology such as 'suspended disbelief' as is found in theatre, film, and literature, one needs to be closely aware of gravity and support between the datasets being linked. Thus, for example if one dataset ends on the right foot with the body crouched forward, the ideal beginning of a linked dataset would be to have the body crouched forward, on the right foot, and facing the same direction.

One theory is of creating a motion library whereby certain sequences of mocap datasets can provide a set of elements that someone can put together to form new combinations. Ballet is often seen this way, where one movement may start from the 5th position and also end in the 5th position. So a sequence such as chaussé, assemblé entornant, glissade, sissonne, tour en l'aire could be reassembled in any combination (there would be 5x4x3x2=120 possibilities).

There are two possibilities for the nature of the datasets. One is that the mocap datasets can all be from the same mover but rearranged.

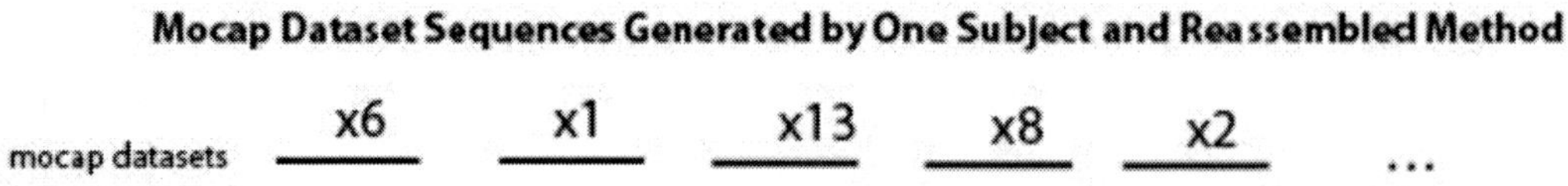

Figure 38. Horizontal sequencing of datasets of one subject.

They can also be culled from different motion producers/talents/subjects.

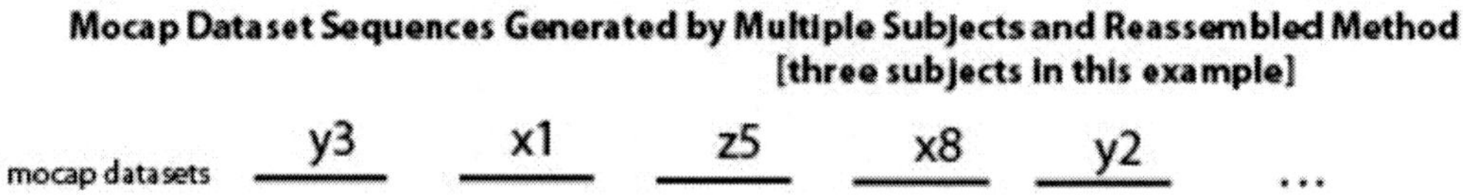

Figure 39. Horizontal sequencing of datasets of multiple subjects.

Blending Mocap Datasets in a Vertical Manner

Consider that you have mocap data in which you like what the legs are doing, but you don't like what the arms are doing; well you can put the legs of one sequence of datasets together with the arms of another sequence of datasets to produce a new dataset that was not originally produced in a biomechanical natural manner.

Similar to the conceptual aspect of blending in a horizontal manner, one can use datasets of the same motion creator, or, of different ones.

Moreover, one can use non-mocap data as part such vertical blending. A special case of this is using mocap data with distortion of figure.

A good thing about mocap data is that those systems that only report x,y.z data for joints (3 degrees of freedom) because they are passive (e.g. optical) can have that data analyzed to create rotational data (6 degrees of freedom).

What this means is that the digital figure does not need to have the same proportions as the talent providing the movement. In other words, someone aesthetically inclined working with mocap data does not need to view the x,y,z numbers of a sequence of datasets to be directly scalable. Rather, given the established joints of the model, the motion can be mapped not as absolutes, rather as relatives.

In *Mercury's Younger Brother*, the expressive idea is that each of us functions according to our 'natural gifts'. While Mercury as tradition and mythology has conveyed, received wings on his ankles to allow him to speedily go from one place to another, we all know that within a family (now, as sentient human beings), not everyone metaphorically receives wings. Sometimes the gifted one is the eldest, sometimes, a middle, sometimes, the youngest. The performance animation expresses a message about doing one's best regardless of the circumstances, and they might be a handicap. [16]

A more striking example of this technique is seen in *Armed with Reach*. I was thinking of empowerment of women (I have two beautiful daughters). I was thinking of a person being active and reaching out for what one desires, not being passive; so I mapped mocap data on a figure and made its arm grow at various moments.

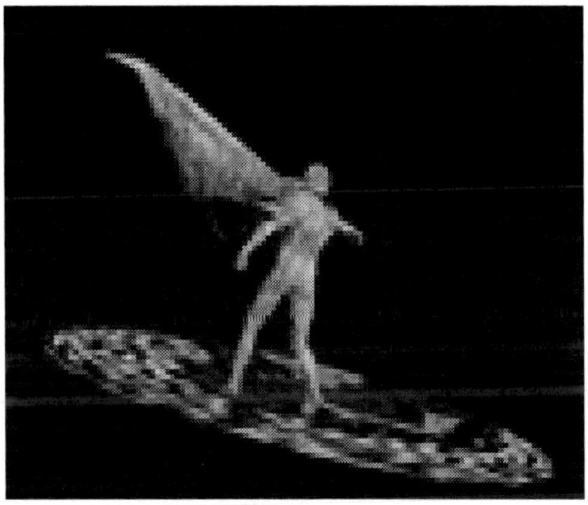

Figure 40. Mercury's Younger Brother (2002).

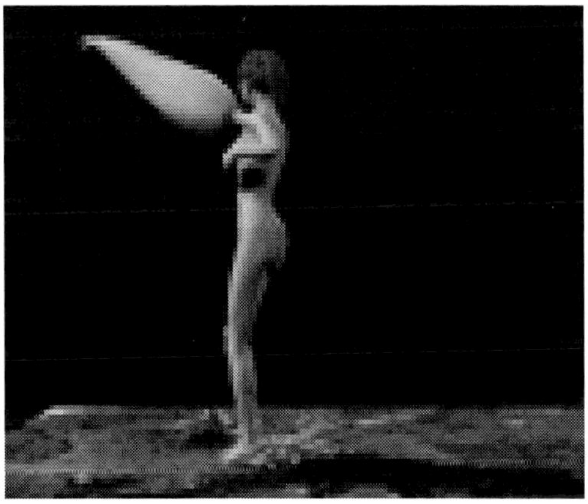

Figure 41. Armed With Reach (2002).

I normally use mocap data I created, with the exception of four pieces - *Ritual at the Well* (my movement data with that of members of my mocap class), *UnderTheSun,OverTheLand* (with movement by Tailfeathers), *Blue____onTheBlackSpan* (my movement data of the figures on the rock, with the jump and reaching movements of the stone figures by Scott Kovacs), and *Broken Egg* (with movement by unknowns, but freely available).

Cross-Mapping Mocap Data to Digital Joints/Bodies

While the normal method of using mocap datasets is a one-to-one mapping from the natural biomechanical joints to those of the digital model, for instance, mapping the mocap data of a right knee to the digital model right knee, one can assign, for instance the mocap right knee data to the digital model left elbow.

I did something like this in one segment of *Broken Egg*. [17]

It is appropriate to mention at this point that mocap data can be cross-mapped to non humanoid or non biped digital objects.

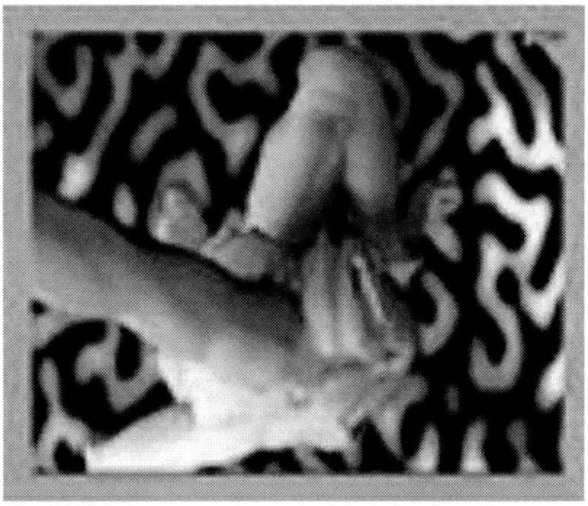

Figure 34. Broken Egg (2002).

Deterministic and Random Techniques of Ordering Datasets

At this point, the reader probably has a clear understanding of the complexity of using mocap datasets to create a memorable expression in the form of an animation. The two extremes of this process are whereby 1) a creator controls as much of the parameters and decision-making as possible and 2) a creator cedes control and accepts the consequences whereby the final product may have some chance, or random aspects to its composition.

When one collaborates, one cedes total control. In effect one is accepting other decision-making as an aspect that is part of the whole.

Yet, regarding movement, there are some ways for a creator to accept experiments that could be successful. For instance, some computer programs allow a movement creator to specify what movements will be blended, what range that blending can occur in, and the probability that such a blend should occur.

Return to Other Time-Based Arts and Their Methodologies

In *musique concrète* of a period fifty years ago or more, a creator takes physical segments of tape and reassembles them. "*Musique concrète* begins with raw sound material, which has been recorded on tape via a microphone. The identity of the sounds or sound objects is transformed by a variety of means and is recorded." [18] Today, a movement creator/expressor/choreographer/assembler can, with mocap datasets, do similar transformations, as has been described and illustrated above.

By the use of the modifier *concrète*, the codifiers of the *musique concrète* believed in an importance of having objective sound data they felt existed naturally and could be captured through their technological intervention, and, then, through manipulation, be altered from its original state as part of human expression.

While *mouvement concrète* suggests backward thinking since it utilizes/recycles terminology, it is an apt way of viewing human organizational practices with natural movement recorded by electronic means. The struggle one has with recorded motion is similar to that of recorded sound decades earlier. Both have data that is additive, and any manipulation of the source material has its effects expressed over time.

Primary, Secondary, Basic or Pure, or Applied Research

The ideas expressed in this article result from primary and basic research. I have always believed in the maxim "The Answers You Receive Depend on the Questions You Ask." What is one person's question might not be that of another. The research products (resulting animations) using mocap datasets are answers to my questions. And if the animations are shared in festivals and evoke thought or aesthetic appreciate in others, they are applied research.

Conclusion

Mocap datasets are being generated by many people with different applications. There are libraries of such datasets. Some people are going to play with datasets whether they be from libraries or are created by talent/movers they direct. Some people like to play, because they are curious. Some people understand, for instance in drawing or coloring, that their artistic representation is not a shared reality based on natural laws, rather it is a personal mapping of elements that really are shared, but are represented using symbolic or reductionist techniques. The same is true with motion. In my opinion, humans do motion best in reality. An animation of human movement can be seen as an imitation, or, if the movement is tweaked, a perception of reality that is felt or imagined, but not necessarily shared in objective time and space. That is all it can be, an imitation, or expression. In the digital representation, we use symbols. Symbols are reductionist by nature.

I have asked myself, why would I want to imitate reality when reality does what it does very well? I don't want to imitate reality. For the most part, reality is banal. Let reality be reality. I want to exaggerate certain aspects of movement reality in the same manner a writer wants to find word symbols to create a picture in your mind about their story. Who can define what story is? It is not just plot. I propose that a concept similar to story exists in all serious mocap animations. I am not prepared to call story narrative and I use the word serious only as a conditional for those animations created with thought and awareness of a beginning, middle, and end, and the awareness of the rightful unfolding of new information. Maybe this is pretentious.

The general understanding of a good film is that if a section is boring, it can't be boring for long. I watch my works and I wonder if they are boring to others. They are not for me. I see a development. Maybe it is the way I have framed the movement from the viewpoint of the camera combined with other changes (for example, the morphing of the body or the speed of revolution of a flattened spherical shape with bold textures).

Mapping objects that don't have temporality is understandable to most people. Art practices often have such objects. They are there; there for a long time. They are tangible. They occupy both time and space. Mapping motion that does have temporality is, . . . well, less easy to seize upon in the mind regarding memory or imagination. It is, at the end - intangible. It is at one part of space at one moment of time. Then it is not there.

In this article, we have explored some methods by which mocap datasets can be manipulated. These techniques are based on experience. There are certainly others, but this survey can serve as a core, or good starting point for those wishing to explore mocap dataset manipulation.

REFERENCES

[1] Barry Schrader. *Introduction to Electro-Acoustic Music*. Prentice-Hall, Inc., Englewood Cliffs, N.J. 1982 p.14.

[2] Gongbing Shan, Peter Visentin, Tanya Harnett. *"A Novel Use of 3D Motion Capture: Creating Conceptual Links between Technology and Representation of Human Gesture in the Visual Arts"* Leonardo Vol 43 No 1 (Feb 2010) 34-42.

[3] Alberto Menache. *Understanding Motion Capture for Computer Animation and Video Games*. Morgan Kaufmann Publishers San Francisco 2000.

[4] To see *ThreeBlueRocks* [though small] in its entirety, go to http://people.uleth.ca /~aw.smith/isabelle.html. I have a theory regarding camera usage. While contemporary perceptions of camera cutting are based on works that require such cutting, for instance T.V. sitcoms and feature films, a non-narrative work featuring movement does not require cutting. It can have cutting, but does not require it - as there are no physical production limitations as in the real world. My personal viewpoint that has guided my decisions for a decade is that if the action/movement is interesting enough, one does not need to add excitement by camera movement.

[5] To see the complete *Dialogos*, go to: http://people.uleth.ca/~aw.smith/digital Choreography.html

[6] To see samples of compression as demonstrated in *Glasshouse* and section 4 of *Journey*, go to http://www.youtube.com/watch?v=RjP7lb8yHgw

[7] To see a sample of amplification as demonstrated in *Ritual at the Well*, go to http://www.youtube.com/watch?v=RjP7lb8yHgw

[8] To see a sample of progression as demonstrated in *3 Marys, RGB*, go to http://www.youtube.com/watch?v=RjP7lb8yHgw

[9] Susan Kozel. Closer Performance, technologies, phenomenology. Cambridge, MA: The MIT Press, 2008.

[10] To see a sample of reversal as demonstrated in the distant figure in *Virilian Hero Dance*, and for a superimposition of five figures [occupying simultaneous space] in *storyTelller*, go to http://www.youtube.com/watch?v=RjP7lb8yHgw.

[11] To see a sample of snipping to discontinuity as demonstrated in *3 Marys, RGB*, go to http://www.youtube.com/watch?v=RjP7lb8yHgw.

[12] To see a sample of snipping to sequencing non-mocap data with mocap data through keyframing as demonstrated in *Purity*, go to http://www.youtube.com/ watch?v=RjP7lb8yHgw

[13] To see a sample of repurposing mocap datasets as demonstrated in *Broken Egg*, go to http://www.youtube.com/watch?v=RjP7lb8yHgw

[14] To see a sample of local gravity referents non-aligned with the world gravity referent as demonstrated in *Non-Plain Planes*, go to http://www.youtube.com/watch?v=RjP7lb8yHgw.

[15] To see a sample of letting mocap datasets be layered in the same space as demonstrated in *Arcadian Women*, and as demonstrated in section 12 of *Journey*, go to http://www.youtube.com/watch?v=RjP7lb8yHgw

[16] To see a sample of vertical layering of mocap datasets with non-mocap data in some joints as demonstrated in *Mercury's Younger Brother*, go to http://www.youtube.com/watch?v=RjP7lb8yHgw

[17] To see a sample of crossmapping mocap data as demonstrated in *Broken Egg*, go to http://www.youtube.com/watch?v=RjP7lb8yHgw

[18] Jon. H. Appleton and Ronald C. Perera. *The Development and Practice of Electronic Music*. Englewood Cliffs, N.J.: Prentice-Hall, 1975. P 71.

IV – Reviews

In: Applications in Arts Biomechanics
Editors: Gongbing Shan and Peter Visentin

ISBN: 978-1-62808-015-5

Chapter 10

APPLICATIONS OF EMG PERTAINING TO MUSIC PERFORMANCE - A REVIEW

Peter Visentin[1] and Gongbing Shan[2]
[1]Department of Music
[2]Department of Kinesiology
University of Lethbridge, Canada

ABSTRACT

EMG is a well proven technology in the fields of motor control and motor learning. Performance mastery on a musical instrument requires some of the most sophisticated neurological processes of any human activity (including rapid interlacing of the cortices that control the auditory, visual, and motor sensory systems). Surprisingly the number of EMG studies in music is small. This review article examines the literature from 1973 to 2010, summarizing research trends in three main areas: 1) the potential of EMG in skill analyses, 2) biofeedback in teaching and learning, and 3) applications in injury remediation and prevention. The aim of this article is to both demonstrate the utility of EMG in music performance research and, in doing so, encourage more research in this area.

INTRODUCTION

Applications of EMG technology in the performing arts vary widely depending on the specific discipline. Artists such as dancers or circus performers can profit and, in some instances have benefited, greatly from EMG research directed at their specific vocational needs. Because such artists' motor skills most closely resemble those of "power" athletes where muscle activity is mainly governed by gross motor control patterns, it is easy to adapt mature methodology from sports into the service of these disciplines (Chatfield, 2003). On the other hand, the use of EMG in performance arts such as music, where fine motor control dominates the motor mechanisms, has been more limited. In these cases, the utility of EMG seems less apparent to both practicing artists and researchers. The focus of this paper is to examine EMG applications in music performance.

In spite of meager amounts of EMG research specific to musicians and music performance, there appears to be significant room for EMG to benefit artists and, by extension, their audience. Through scientific research, artists can potentially learn to acquire skills more efficiently and effectively, speeding up the learning process and contributing to more technically reliable performances. Further, concerns about vocational injuries has become a significant issue in the performing arts and cross-disciplinary research involving scientific methodologies can prove to be powerful influences in areas of both prevention and remediation (Chesky, Kondraske, Henoch, Hipple, and Rubin, 2002; Lehmann and Davidson, 2002; Visentin, Shan, and Schultz, 2004; Wilson, 1986). In this regard, some technologies and methodologies from sports science have already been successfully adapted to music research (Shan and Visentin, 2003; Visentin and Shan, 2003). While EMG research in sports has been an acknowledged aid in improving sport performance for decades (for example, the Journal of Electromyography and Kinesiology has existed for more than 20 years), the state of affairs in 2011 remains similar to that expressed by Basmajian and White in the journal *Nature*: "Scientific experiments on musical performance are surprisingly sparse when one considers that neuromuscular control of skilful motor performance reaches its acme in music" (Basmajian and White, 1973). The primary barriers to EMG research in music seem to be both philosophical and practical.

Generally, music performance is rooted in training methodologies governed by aural tradition, with pedagogical frameworks by and large limited by conventions and aesthetic values passed from teacher to pupil. Participants in such a model typically take great pride in the provenance of their traditions – often validating their artistic values by tracing their musical lineage from teacher to teacher, to one or another of the legendary "greats" of the past. Such a mind-set frequently inspires a mystical and, in some cases, a near-religious reverence of the teacher, the traditions, the aesthetic concepts, and more importantly "technique". Philosophically to some degree, music, even in its most heavily documented traditions (e.g. western classical music), remains an arcane art firmly rooted in centuries of convention.

In Western classic traditions, most recent developments in pedagogy can be understood as a general move from a collective consciousness to an aesthetic with a focus on individual virtuosity. Romantic writers of the 19th century typically viewed virtuosic technique as something superhuman, transcendent of physical limitations (Kolneder, 1993; Metzner, 1998; Samson, 2003).

On one hand, teaching becomes an arcane art both mysterious and incomprehensible, and on the other hand, the period marks the beginnings of music Pedagogy as a Science. Contemporary writing emphasized both "genius" and the development of technique as a means to achieve the highest degrees of artistry (Samson, 2003). "Schools" of teaching arose based on an individual teacher's influence or on a unique aspect of playing that was deliberately cultivated by the school's followers (such as use of vibrato).

In order to challenge the inertia of entrenched traditions, researchers and research methods must have a convincing raison d'être before they will gain any currency in the music industry. As a technology, EMG still needs to prove its utility in this regard.

Practical barriers to implementation of EMG methods in music research are also significant. Unlike sports or other aerobic activities relying predominantly on gross motor control dominated by large muscles, much music performance is dominated by fine motor skills involving small muscles that are often also deep ones (for example the flexor policus longus, extensor digitorium and Interosseous I).

Since surface EMG (sEMG) records any and all seizure potentials released by the muscles over the area of and between the sensors, the possibility of interference (cross-talk) of neighbour muscles increases dramatically when measuring small, deep and/or bunched muscle groups. One possible solution for such challenges might be to employ fine wire techniques. However, there is a significant concern that this technique has high potential to influence the outcomes of performance, effectively negating some of the research benefits that might be achieved.

The simple fact of the matter is that artistic success or failure is often determined by micro-increments of skill, and research results where a potential influence on the outcome is high would likely be viewed by artists as having little practical value.

A further barrier to using sEMG in music research revolves around difficulties in interpreting resultant data. Unlike most applications in sports where larger muscles such as the biceps or quadriceps are targeted, the goals of music research necessarily include both large and smaller muscles such as biceps and triceps in the upper arm, and flexor pollicis brevis and adductor pollicis in the hands. Hence, results will include some data that is easy to examine using well-developed sports methodologies, and some data that is much more difficult to interpret.

One of the most effective uses of sEMG in pedagogy/training involves using measured signals to fractionate the body's response process in order to get timing of neural control as well as the coordination among muscle groups. Small muscles and current configurations of sEMG sensor technology make it very difficult to decipher the dissimulation effect of cross-talk. Existing sEMG technology and data analysis methods need to be further adapted to accommodate the differing conditions and demands of music research.

Since existing research employs, in greatest proportion, surface techniques, the remainder of this review article will use "EMG" to refer to surface electromyography. The article will sample existing the literature where EMG is employed in music research, examining results, their potential significance and discussing research challenges in the areas of pedagogy, injury prevention and technological limitations. In pedagogy, discussions will include skill analysis, biofeedback training, and the evaluation of training methods. In the area of injury research, exemplary remediation and prevention uses of the technology will be presented. Finally, challenges to the effective use of the technology and their importance to ongoing research needs will provide a point of departure for speculation on the future of EMG research in the field of music.

In short, EMG research in music finds itself at a cross-road. Technological and analytical limitations, meaningful research results, and cultural buy-in from the music industry are interdependent.

Currently, the existing body of research is below a critical mass needed to encourage growth in the area. Interesting and significant research involving innovative uses of EMG, it is hoped, can be a catalyst for encouraging new researchers to become involved in music applications and for industry to continue to improve the technology in ways that might not be demanded not only in music but also in other research areas. Further, relevant research results can be a way to show utility of the technology to musicians, especially in the service of pedagogy and injury prevention.

Such work can improve the lot of musicians as a whole by changing attitudes in industry. Just as EMG has become a significant scientific tool in sports research, the authors propose that similar, but as of yet underexplored, potential exists in the music research.

Application Potential in Skill Analyses

EMG provides a non-invasive means to gather data about physiological processes that cause muscles to produce movement. Muscle activity is reflected in seizure potentials measured by EMG, which can then be used to correlate intensity, timing and duration of activity.

Consequently this information supplies insight into response processes, neural control, and coordination among muscle groups. Such analyses permit assessments of motor control and technical proficiency, which collectively provides foundational information for biofeedback training and development of new pedagogical methods. As such, one might expect EMG to be a scientific tool of choice for analyzing musicians' **motor skills** but, perplexingly, the body of EMG research in music pedagogy over the last 30 years is very small (Shan and Visentin, 2011). A review of the literature shows that EMG applications tend to fall mainly into only three broad categories: skill analysis descriptions and establishment of norms, training method evaluations, biofeedback in both pedagogy and injury remediation (Kjelland, 2000).

In the areas of skill analysis and evaluation of training, methods in music research are typically borrowed from human movement sciences, where EMG is used to quantify motor control and development, coordination and skill optimization (Riley and Chong, 2010). To do so researchers must process raw data since EMG signals are dependent on individual biological make-up such as skin thickness, subcutaneous layers of fat or other soft tissues, and placement of the sensors.

Thus raw signals are unavoidably *non-quantitative* measures of neural-muscular activity (e.g. a sumo wrestler might generate lower measured signals than a 90-pound ballerina, yet no one would be surprised at the outcome of a basho match-up between them). Processing methods and analysis techniques play an vital role (Shan and Visentin, 2011).

Typically two processing steps are used in order to provide *quasi-quantification* of raw data. Enveloping stabilizes stochastic raw signals. Normalization of enveloped data expresses all measurements as a percentage of a chosen reference value (Kjelland, 2000; Morasky, Reynolds, and Clarke, 1981; Zinn, 1998). In this way, EMG can provide a viable means of comparison among different subjects and trials. Commonly used normalization references include: 1) the maximum signal level during a trial, or 2) a Maximum Voluntary Contraction (MVC) determined in a separate trial under high-load conditions.

Skill analyses using EMG have been undertaken in multiple areas of vocal instrumental music performance (piano, winds and strings). Below are representative examples of skill analyses in each of these areas.

Voice Skill Analyses

One example of music skill analysis in the existing literature looks at the skill of singing onsets (Shan and Visentin, 2010). Onset is a music term used for the articulatory beginning of the sounding of a pitch, whether vocal or instrumental. As such, it is a vital component in musical expression, and the effort used to initiate a tone, if it can be accurately measured, should show something about technical proficiency of the musician. Since consistency of timing and minimization of effort can be used to measure motor control, analysis of onsets can illuminate preparation and execution during singing or playing. This, in turn, can prove or disprove empirical evidence of traditional learning and teaching.

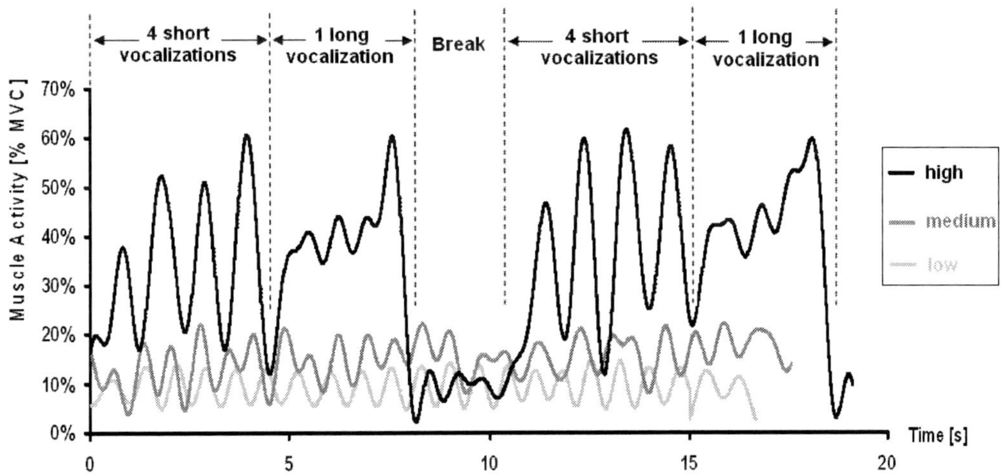

Figure 1. Rectus abdominis activity of a trained mezzo-soprano at three different pitch frequencies (modified from Shan and Visentin, 2010).

Figure 1 shows rectus abdominis activity of a trained mezzo-soprano singer executing onsets at three different pitch levels. It shows ten onsets grouped into two sets of five with a short break in between sets. Each set shows four short vocalizations (approx. 1s. each) followed by one long vocalization (approx. 4s.). Three different pitch frequencies were tested: high, middle, and low. The black line shows EMG activity for the high pitch (F5, or two Fs above middle C on the piano). Dark grey and light grey represent medium and low respectively (or F4, just above middle C on the piano, and F3, just below Middle C).

This kind of analysis can be used to evaluate the effects of training. It is normally expected that, for training *in any* kind of activity involving motor control, stability is a measure of ability; novices will have a less consistent pattern than experts. The consistency and repeatability exhibited in Figure 1 shows the subject to have a high level of competency. Additionally, varying amplitudes at different pitch levels also show us something. Significantly different amounts of physical effort are required at different pitch levels. At high pitch, peak levels of EMG were above 60% of the MVC with breaks between short vocalizations around 15%-25% of MVC and the longer break between sets under 10% of MVC. For medium and low pitches peak levels, EMG was significantly lower and there was no insignificant difference in EMG data between singing and rest (all ranging between 5%-20% of MVC). A non-linear relationship between pitch and effort is shown. For this singer, control was very consistent and, especially from EMG data of the high pitch, the disciplined preparation and execution is typical of individuals that are highly skilled. In the hands of a knowledgeable and skilled teacher, such results have implications for teaching and learning because it provides biofeedback information that can help improve training methods and technique acquisition.

In an earlier EMG study of voice, Englehart and his colleague examined the relationship between vocal tone quality and preparatory psychomotor sets (Engelhart and Major, 1989) . His concern was that existing research had overlooked the issue of preparation, rather concentrating on tone quality post phonation. A second thrust of his dissertation was to examine how training using the Alexander technique (which focuses on body alignment and posture) could influence outcomes. He found that training using EMG and the Alexander technique could positively influence muscular tensions in the neck and upper back, particularly in the upper trapezius muscle. Further he found anticipatory muscle activity changed significantly on higher pitches but not in response to increased dynamic levels beyond forte. Because limitations of the study were significant, some of his initial hypotheses could not be proven. However, this research is important since it initiated the use EMG for the purpose of skill analyses in relation to specific pedagogical techniques associated with relaxation and vocal production.

Piano Skill Analyses

In a literature review published in 2000, Kjelland found that the majority of studies since 1985 employed EMG as a descriptive/diagnostic tool for the purposes of establishing performance norms (i.e. skill analysis), comparing teaching methods and techniques, and diagnosing performance difficulties (Kjelland, 2000). A 1989 study used EMG to compare three different technical approaches to playing the piano (F. Bejjani et al., 1989). In this case study, a single player was asked to perform three pianistic tasks from a single composition, using three defined postural orientations of the hand. Although it may appear to have been somewhat artificial to have the same player adopt multiple postural orientations, rather than simply using his/her normal one(s), the study validated the utility of EMG as a means to discriminate technique-related differential muscle activity in pianists. In another study, Montes and his colleagues summarised some interesting skill analysis applications where the EMG activity in forearm and elbow were examine for pianists at different levels of proficiency (Montes, Bedmar, and Sol Martin, 1993). Muscular patterns of activity establishing skill norms were defined using the data from advanced pianists. Phases of EMG activity examined included preparatory key attack (before piano keys were pressed), key rest (lowest levels of EMG activity) and key holding (where EMG varied according to the attack strength). These characteristics were not observed in beginners; rather, the EMG signal showed an overall tension to be present in the muscles. These studies can be used as exemplars of the first necessary steps in any EMG application – i.e. targeting specific phenomena and evaluating the characteristics of the activity.

Skill analyses prove to be a first step for other piano studies where EMG was used as part of an effort to integrate multiple technologies for use in retraining of complex technical skills (Riley, Coons, and Marcarian, 2005). Such work builds on results from previous EMG studies where professional pianists were examined and it was determined that, after depressing a key, they seem to relax immediately (Montes, et al., 1993; Zinn, 1998). Contrarily, in beginners and novices a truncated tension-relaxation cycle was identified as a problem.

In a more detailed skill analysis study, EMG and strain gauge were used to examine whether pedagogical training, a potential cause of structural adaptations in hand intrinsic muscles, was aerobic or anaerobic by nature. Changes in EMG power spectrum during incremental isometric muscle contractions were studied to evaluate the effects of training using pianists and an untrained control group (Penn, Chuang, Chan, and Hsu, 1999). Although the MVC was similar in both groups, the effects of training resulted in pianist having increased endurance as measured by time to induced fatigue. The authors of this study suggested results to further indicate that piano training should be classified, endurance training rather than power training. Such evaluations of skills using EMG can identify muscle recruitment strategies and provide information about fine motor training for music learners and professional and physical adaptations (differences in muscle fibre composition) because of training. In piano performance, it is generally accepted among researchers and performers that that sustained levels of muscle contraction inhibit smooth motor movements and contribute to fatigue in both practicing and performing. EMG provides an excellent tool for measuring these characteristics during human movement.

Brass/Winds Skill Analyses

In wind instrument performance, it is also generally agreed that unnecessary muscle tension not only leads to physical problems but also can interfere with performance quality (Morasky, Reynolds, and Sowell, 1983). In this study, EMG was used to examine muscle tension in eight intermediate to advanced clarinet players where skill analysis assessed short-term and extended generalizations of EMG levels along with trill and scale speed scores. In a series of four other studies from 1991 to 1998, Heuser and McNitt-Gray undertook skill analyses associated with trumpet embouchure and tone production. These linked studies go well beyond the merely descriptive. Skill analyses were contextually correlated to articulatory attack (1991) by examining EMG potentials as the performers prepared for the commencement of playing a tone. Results indicated that EMG was an effective way to identify some aspects of tone production difficulties. In the second study (1993), the researchers examined asymmetrical mouthpiece placement. Asymmetry is frequently a compensatory mechanism used by trumpet players where embouchure placements differ from the theoretical "centred" ideal to accommodate oral-facial make-up variance or other technical difficulties.

The study compared professional trumpet players who use asymmetrical placement of the trumpet mouthpiece with a sample of players who perform with the mouthpiece centred on the embouchure. Although results indicated that there was no significant difference in EMG readings between the two groups, they are meaningful none-the-less.

The data allows us to induce that the compensatory behaviours being adopted by those who played asymmetrically were indeed successful for (at least for this one aspect of the motor mechanism – muscular effort) overcoming challenges associated with oral-facial variance. The question remains whether such compensation is a good long-term strategy for overcoming "technical deficiencies".

The third and fourth of the Heuser and McNitt studies (1994 and1998) draw EMG applications into the realm of pedagogy, which will be examined later in this paper. Together, these studies show a progression from description to application of EMG research in music performance. A more recent EMG skill analysis study for brass players was undertaken in 2005 (Iltis and Givens, 2005). The purpose of the study was to use EMG to quantify the activity of selected facial muscles during French horn playing, comparing a performer with distonic facial muscles to a group of unimpaired players. The findings suggest that a relatively simple EMG testing protocol could be helpful in both skill analyses and comparative research for injury prevention/remediation.

Strings Skill Analyses

In string performance, EMG technology has been used to observe muscle activation patterns and levels as well as making some comparisons to establish norms of musculoskeletal functioning during performance. In a 1988 study, Moore et. al. looked at cellists performing trills in order to determine the characteristics and timing of the finger movement of and the associated of EMG activity of muscles that control the fingers (Moore, Hary, and Naill, 1988). Such information is foundational for generating research questions that permit training method evaluations and biofeedback studies of performance-related problems. A 1992 study examined one of the most fundamental questions for violinists and violists – the effects of using a shoulder rest to improve the position of the instrument and overall posture (Levy, Lee, Brandfonbrener, Press, and Levy, 1992). EMG was used to evaluate Trapezius and sternocleidomastoid muscles (neck and upper back) and findings revealed that use of a shoulder rest had significant implications for the muscles that supported the holding of a violin. Weber investigated EMG muscle potentials during vibrato motion of violin and viola performers during the playing of standardized exercises (Weber, 1995). Results demonstrated consistent firing patterns and found that forearm muscles appeared to be the prime motivators of the vibrato technique. In another violin vibrato study, Bejjani et al. used EMG to study fourteen violinists. The results were analyzed using qualitative assessment of the EMG signals and correlation of muscle activity with vibrato motion and time/frequency parameters (F. Bejjani, Ferrara, and Pavlidis, 1989). Such descriptive work leads to the establishment of norms which set objective and reproducible parameters of musculoskeletal functioning during the performance of musical tasks (Kjelland, 2000).

SUMMARY

In music, EMG applications must begin by undertaking detailed analyses of specific skills. Necessarily there needs to be a body of descriptive research that sets foundational groundwork for research of a more complex nature. Skill analyses of a descriptive nature can be used to establish norms, bases for training method evaluations, and development of new training methods based on a biofeedback approach in both pedagogy and injury remediation. Although the existing body of literature is small, it demonstrates the substantial utility of EMG technology. It shows forward-looking thinking, as researchers appear to have carefully considered both 1) what questions to ask and 2) what issues of greater significance these questions underpin. The authors wonder why the body of research in this area remains so small when demonstrated need, particularly in the area of improving pedagogy, is so apparent. We can only speculate that the necessarily cross-disciplinary nature of application-oriented research in the arts is inhibited by the number and variety of skill sets that need to be brought to the research table. Unlike sports and kinesiology where gross motor control and power dominate applications of the technology, the kinds of fine motor control typically involved in performance pose their own challenges to the capabilities of EMG technology. Further, EMG research in music requires a meeting of the scientific mind with an artistic and creative one that typically eschews definitions and boundaries. Individuality is a part of the creative process and researchers must carefully design their work to negotiate a path where both science and art are equally part of the underlying motivation of the research.

Biofeedback Applications in Music

Biofeedback techniques may help musical performers in a number of ways: 1) promoting general relaxation, 2) helping to reduce performance anxiety, 3) aiding in the treatment of functional disorders, and 4) improving pedagogy and training for musicians. In motor learning, the use of EMG biofeedback as a means to speed up the training process is common. Stress minimization (both physiological and mental) is a valuable function of biofeedback training (French, 1980; Niemann, Pratt, and Maughan, 1993). In music, training using EMG as a tool for biofeedback offers multiple potential benefits beyond the mere controlling of excess muscle tension during performance. Importantly, reduced EMG levels as a result of biofeedback training can generalize to situations in which the feedback is not available (Morasky, et al., 1983).

In the identified study, Morasky et al. came to the conclusion that a reduction in muscle tension from biofeedback training did not result in decreased performance quality for specific skills. This is an important finding because it points toward a debunking of any mythos that might be associated with muscular tension as a tool of individuality/expression with regard to skill effectiveness and the creative process.

Biofeedback training can alter basic industrial/workplace preconceptions as well. With the interventions of technology, outcome-oriented thinking should become more process-oriented, and improve both efficiency of skill acquisition in terms of time spent learning and reliability of the skill. Efficiency in learning may have potential benefits for effectiveness of artistic outcomes beyond mere reliability (Visentin, Shan, and Wasiak, 2008). Perhaps most importantly, optimization of skill control is acknowledged to have significant potential in terms of injury prevention (Cummings, Wilson, and Bird, 1984). As a means to speed up the training process in music, while avoiding unnecessary injuries, biofeedback training has an as-of-yet under-explored potential in terms of making skill execution autonomous and freeing musicians' minds to focus on artistic creation (Shan and Visentin, 2010).

Biofeedback studies with direct links to pedagogy have been undertaken in the areas of voice, piano, winds and strings. In spite of initial industrial interest in EMG biofeedback as an aid to pedagogy, particularly in the voice area, there are surprisingly few scientifically credible studies. Similarly, for piano, the body of literature is exceedingly small. In the winds and strings areas, although still small, the representation of scientific work using EMG is larger perhaps because of the more complex and/or asymmetrical postural behaviours associated with instruments such as violin, double bass, and trumpet.

EMG research in terms of biofeedback training for musicians builds on the general theories of the late 1970s and increases to its peak level of activity in the 1980s (French, 1978, 1980; Garrison, 1978; Irvine and LeVine, 1981; Levee, Cohen, and Rickles, 1976; LeVine, 1983; LeVine and Irvine, 1984; Moore, et al., 1988; Morasky, et al., 1981; Morasky, et al., 1983). Over the last 20 years, an unexpectedly small number of biofeedback training studies are found, and those between 1990 and 2000 tend to follow the models and methods of the 1980s (Heuser and McNitt-Gray, 1998; Montes, et al., 1993; Niemann, et al., 1993). After 2000, studies are fewer still, even though new developments in technology make possible the use of multi-modal or multidimensional signal analysis studies (Visentin, et al., 2004).

Most recently EMG has been effectively incorporated in multimodal biofeedback studies using aural (Disklavier recording) and visual (video and computer-based output) feedback, motion analysis and EMG. Measured signals have been used in both real-time feedback pedagogy as well as data for post-process analyses (Riley, et al., 2005; Yoshie, Kudo, and Ohtsuki, 2008). It remains to be seen if the use of multiple biofeedback sources will improve the effects of training more than well-selected use of individual technologies.

Voice and Biofeedback Training

One of the first contextual uses of EMG for biofeedback training is in the area of voice instruction (Garrison, 1978). His doctoral dissertation looked at problems encountered by students with tightness in the throat and jaw muscles. The study aimed to supply a comparison between students who only received traditional vocal instruction and those who received supplemental biofeedback training using EMG to help overcome tensions or tightness in identified muscles. Subjects of this study were college voice students suffering from tension problems. The results showed that there was a significant difference in "improvement" between those singers with tension problems who received EMG biofeedback training and those who did not.

Piano and Biofeedback Training

In the area of piano teaching, an early study by Montes et al. used EMG biofeedback as a supplemental method to piano teaching. Examining the abductor pollicis brevis as an agonist during selective movements of piano playing, the researchers found an increase in the peak amplitude and the relaxation values for a group using biofeedback compared to a control group. Such results have clear implications for the application EMG biofeedback as an aid in piano pedagogy (Montes, et al., 1993). Similarly, a 1998 thesis also explored EMG biofeedback as a means to enhance relaxation into piano performance (Zinn, 1998).

In a 2005 study, Riley worked to improve piano technique through multi-modal biofeedback training focusing on proper skeletal alignment of hands, arms, and shoulders. The goal was to avoid repetitive practice (typical of student learners) where poor balance between muscles activity and intended outcome might lead to "difficulties with, rather than mastery of, technique and stylistic interpretation and even physical injury" (Riley, et al., 2005).

This new approach utilizes multiple technologies, including EMG, as objective assessors of technical elements traditionally considered to be the subjective purview of a music instructor. As such, it was speculated that learning could become a more interactive process between teacher and pupil because an experience mediated by technology makes both aware of problems in technique and of ways to correct them. The concept of multi-modal feedback as a tool in pedagogy was extended by Yoshi et al. in 2008. In a study examining performance stress and anxiety, EMG was used along with measurements of heart rate, moisture (sweat), and MIDI sound signals. The main findings suggested that psychological stress could increase the risk of playing-related musculoskeletal disorders (Yoshie, et al., 2008). Such results provide convincing rationale for the use of EMG as one element in multi-modal pedagogical applications dealing with both physical and psychological challenges faced by learners and performers.

Brass/Winds and Biofeedback Training

In brass/wind performance, representative EMG biofeedback studies have been undertaken for several different instruments (Heuser and Mcnitt-Gray, 1994, 1998; Levee, et al., 1976; Morasky, et al., 1983). Since the mechanism of sound production involves the performer blowing into the instrument, most of these studies concentrate on facial and/or throat musculature. The general aims are to explore the minimization of tensions in order to improve performance ability. In the Levee study, biofeedback relaxation training was focused on muscles where EMG showed chronically high tension levels. Results showed biofeedback training to dramatically help reductions in tension levels. Further, it appeared that there was an increase in the patient's proficiency with the adoption of new motor control patterns that could be used in his professional activities (Levee, et al., 1976). Such biofeedback training can have both physiological and psychological benefits for the player. Notably, Heuser undertook a series of four studies, where the first two were skill analyses (previously discussed) and the last two of which used results obtained to underpin biofeedback training (Heuser and Mcnitt-Gray, 1994, 1998). The 1994 study examined tone commencement (onset articulation) using EMG to measure the training effect on embouchure muscles for learners encountering tone commencement difficulties. The second was a case study of a student experiencing both playing difficulties and pain. It used real-time EMG feedback to provide visual substantiation of improvements related to the learning strategies designed by the instructor. Typical of EMG biofeedback training, these studies employed "1) visual demonstration of how changes in performance strategies relate to changes in muscle activation patterns, and 2) ... visual confirmation of the contribution specific instructions make toward restructuring muscle activation patterns." (Heuser and McNitt-Gray, 1998). All of these studies show that EMG can be an effective tool in biofeedback training by supplementing traditional teaching methodologies and validating pedagogical practices with an objective means.

Strings and Biofeedback Training

EMG studies for string players are more varied than those for any other instrumental/vocal group. The asymmetrical nature of playing the instruments and the awkward postural orientation of violin and viola in particular is a recipe for creating unwanted muscular tensions, something that EMG technology is well suited to explore. Further, the complexity is compounded by the required sensitivity of motor control, making skill optimization a central concern for all players. This is best summarized in the pedagogical writings of Kato Havas, an internationally recognized teacher who focussed on optimizing the mechanics of violin playing in a unique and effective manner; "Playing the violin is never difficult; it is either easy or it is impossible." (Havas, 1961). Consequently, EMG biofeedback seems an ideal tool for demonstrating muscle activation patterns during skill acquisition and validating pedagogical approaches.

This was first recognized in studies by Morasky (1981) and Levine (1984). Both of these studies aimed to see if EMG could be used to reduce unwanted muscle tensions. Morasky found that biofeedback facilitated significant decreases in EMG of the arm muscles. Levine employed EMG as a pedagogical tool to decrease unwanted left-hand tension for a group of violinists and violists. He noted that improvement was rapid and indeed persisted over a 5-month follow up period (LeVine and Irvine, 1984). Surprisingly, iIn spite of these early successes, EMG is still not widely applied in strings pedagogy.

For lower strings, double bass and cello studies by Guettler (1992) and Theim (1994), respectively, examined left-hand techniques. The Guettler study additionally examined instrument orientation to the performer's body. Vibrato technique on the Bass was found to be is associated with pulsating contractions of two back muscles. A student having problems in producing a proper vibrato was observed to have poor coordination between these muscles (the teres minor and the teres major). A second experiment clearly demonstrated that small changes in positioning the instrument had noteworthy influence on muscular activity in the left and right trapezius muscles (Guettler, 1992). The pedagogical implications of such analyses are foundational for biofeedback work. In the work of Thiem et al., rhythmic cuing (the grouping of several notes together initiated by a single gross-motor-control gesture) was trained over several weeks and evaluated using EMG. Pre- and post-test results revealed the pedagogical strategy to have significant benefits in terms of lowered muscle activity/tension (Thiem, Greene, Perassas, and Thaut, 1994). Bowing techniques have also been the subject of biofeedback training studies using EMG as found in a summary of Koehler's dissertation work exploring two specific bowing techniques (Koehler, 1995). Studies such show the effectiveness of EMG as a pedagogical aid.

Music Performance and General Biofeedback Training

Finally, in addition to instrument-specific biofeedback and pedagogy studies, there is work generalizing the use of EMG as a pedagogical tool across instrumental lines. The use of biofeedback as a means to decrease overall levels of performance anxiety is best exemplified in the work of Niemann (1993) and Reitman (2001). Anxiety-coping strategies were evaluated by using biofeedback training and music relaxation interventions for a group of music students. The work resulted in a significant reduction of performance anxiety

(Niemann, et al., 1993). EMG was used to confirm the effectiveness of anxiety reduction strategies through measurements of reduced muscle activity/tension in the frontalis (Reitman, 2001).

SUMMARY

In terms of biofeedback training in the teaching and learning of music, EMG clearly has significant potential to improve skill optimization and skill effectiveness. What is less apparent is that it can aid the creative process by reducing both the physical and mental demands of acquiring and using technical skills, freeing the artist to focus on the artistic ones. The complex behaviours typically found in music performance most certainly benefit from any training method that speeds up the learning process and does so in a way that provides generalized enhancements in skill development and motor control. EMG has particular relevance in terms of reducing unnecessary or unwanted muscular tensions. Thus, it can also provide a means to minimize the psychological pressures that are understandably associated with live music performance – a high-skill, high-pressure activity that has a zero tolerance for technical errors.

Injury Remediation and Prevention

Musculoskeletal problems among professional musicians are mostly linked to type of work involved in performing on an instrument. Certain body postures and movements are maintained for extended periods resulting in repetitive overuse of the muscles and soft tissues (Hoppmann, 1998; Owen, 1985; Visentin, et al., 2004; Zaza, 1998). The prevalence of playing related musculoskeletal disorders (PRMSDs) is particularly highlighted in epidemiological studies undertaken between 1988 and 1998. In these studies, the rate of injuries for musicians having at least one medical problem severe enough to affect their performance peaks at 76% (Fishbein, Middlestadt, Ottati, Straus, and Ellis, 1988; Middlestadt and Fishbein, 1989; Zaza, 1998). Since, for PRMSDs, it is the very act of "playing" that causes problems, quantification of musculoskeletal stresses during activity could be very important in determining the aetiology of injury. EMG provides a valuable means to begin to differentiate the gestural motivators of the performative act. It can be used to observe and evaluate muscle activation patterns, performance behaviours contributing to PRMSDs, improvements during training or reestablishment of skills post-injury. However, it is important to consider that since performance mastery on a musical instrument requires some of the most sophisticated neurological processes of any human activity (including rapid interlacing of the cortices that control the auditory, visual, and motor sensory systems), injury research in its ultimate manifestation might also need to address more fundamental brain and behaviour phenomena.

Empirical evidence suggests that for fine motor control dominated activities the presence of an injury may necessitate a performer invoking compensatory strategies in order to minimize pain while continuing performing. Applications of EMG that are remedial in nature can help establish the effectiveness of compensatory motor control mechanisms, whether through retraining or medical intervention. In a study of the neck and shoulder area for violin and viola players, EMG activity in the upper trapezius was measured under three circumstances demanding three different levels of effort: rest, performance of an easy piece, and performance of a difficult piece (Berque and Gray, 2002). Although preliminary in nature because of a small subject sample, the results pointed toward potential compensatory strategies, suggesting that injured subjects may have redistributed muscle loading from the measured muscle to other synergistic ones as a strategy to alleviate discomfort. A second representative study goes further (Steinmetz, Ridder, and Reichelt, 2006). It used EMG to confirm the effectiveness of a medical intervention. In violinists, craniomandibular dysfunction is related to increased muscular load in the muscles of mastication, the trapezius and sternocleidomastoid muscles. Steinmetz et al. explored the use of occlusal splints to lower muscle loading in the masseter, temporalis, trapezius, and sternocleidomastoid during violin performance. The main finding was that EMG showed decreased muscular load in symptomatic and asymptomatic violinists, suggesting that splints could have potential roles in both remediation and prevention.

Some of the most intriguing and forward-looking applications of EMG occur in conjunction with electro encephalography (EEG) and, more recently, functional magnetic resonance imaging (fMRI). In this regard, the work of researchers associated with the Institute of Music Physiology and Performing Arts Medicine, Academy of Music and Drama in Hannover, Germany is exemplary.

Among other projects, these researchers have taken on the serious issue of musician's cramp (or focal distonia) through longitudinal studies. Focal distonia (FD) is a task-specific pathophysiology associated with repetitively executed complex movements, such as playing a musical instrument (Altenmüller, 2003; Lederman, 1988). It involves an involuntary cramping of single fingers during the execution of complex neural motor movements that is as completely painless as it is almost inescapably career ending or, at the very least, career altering. Given the seriousness of outcome, there is a growing body of research investigating the essential role of the somatosensory system in the pathophysiology of FD.

In a representative study, Ruiz et al. investigated neural correlates associated with motor memory in pianists with FD using EMG in conjunction with multichannel EEG (Ruiz, Jabusch, and Altenmüller, 2009). In an earlier study, proprioception (unconscious perception of movement and spatial orientation arising from stimuli within the body itself.) investigated the effects of muscle vibration using transcranial magnetic stimulation on motor cortical excitability in FD patients by comparing that of unafflicted musicians (Rosenkranz, Altenmuller, Siggelkowa, and Dengler, 2000).

From the above studies, it is clear that EMG, as a relatively easy to use and non invasive technology, has great potential in the study of musicians' injuries. Perhaps its most important contributions might point toward prevention rather than remediation. In this regard, EMG research needs to incorporate skill analyses and biofeedback techniques as a means to influence pedagogy and industrial attitudes in order to minimize risks of vocational injury among musicians.

CONCLUSION

In conclusion, EMG research in music finds itself at an awkward stage. Unfortunately, the number of existing and published EMG studies related to music is still very small, well below a critical mass needed to clearly highlight the advantages of EMG as a tool in research and teaching. The studies available demonstrate the varied utility of the technology, especially in the service of music pedagogy and injury prevention for performers, but a larger body of research is needed and research methods must be adapted or proven in multiple musical contexts in order to meet musicians' perceptions and expectations regarding the requirements of both the performer and the music industry.

EMG research in music is still at the stage of establishing a basic body of descriptive research, setting foundational groundwork for research of a more complex nature. In order to gain acceptance among musicians, skill analyses, training method evaluations, development of new training methods, and biofeedback studies in both pedagogy and injury remediation need to be application-oriented, a condition that is contingent upon cross-disciplinary research design; thus, research design should have the capacity to negotiate a path where both science and art are equally part of an underlying motivation (i.e. EMG research must have sensitivity to the artistic and creative aspects of the music vocation). Finding common ground with artists will be key. The complex motor behaviours in performance can certainly benefit from efficiencies identified in EMG research. Generalized enhancements in skill development and motor control and avoidance of injuries have particular relevance in today's music industry. It remains for technology and the body of music performance research to mature to a point that encourages self-sustained innovation and growth.

REFERENCES

Altenmüller, E. (2003). Focal dystonia: Advances in brain imaging and understanding of fine motor control in musicians. *Hand Clinics, 19*(3), 523-538.

Basmajian, J., and White, E. (1973). Neuromuscular control of Trumpeters' Lips. *Nature, 241*(Jan 5), 70.

Bejjani, F., Ferrara, L., and Pavlidis, L. (1989). A comparative electromyographic and acoustic analysis of violin vibrato in healthy professional violinists. *Medical Problems of Performing Artists, 4*(4), 168-175.

Bejjani, F., Ferrara, L., Xu, N., Tomaino, C., Pavlidis, L., Wu, J., and Dommerholt, J. (1989). Comparison of three piano techniques as an implementation of a proposed experimental design. *Medical Problems of Performing Artists, 4*(3), 109-113.

Berque, P., and Gray, H. (2002). The influence of neck-shoulder pain of trapezius muscle activity among professional violin and viola players: an electromyographic study. *Medical Problems of Performing Artists, 17*(2), 68-75.

Chatfield, S. (2003). Variability of Electromyographic and Kinematic Measurement in Dance Medicine and Science Research. *Journal of Dance Medicine and Science, 7*(2), 42-48.

Chesky, K., Kondraske, G., Henoch, M., Hipple, J., and Rubin, B. (2002). Musician's Health. In R. Colwell and C. Richardson (Eds.), *The new handbook of research on music teaching and learning*. New York: Oxford University Press.

Cummings, M., Wilson, V., and Bird, E. (1984). Flexibility development in sprinters using EMG biofeedback and relaxation training. *Applied Psychophysiology and Biofeedback, 9*(3), 395-405.

Engelhart, R. J., and Major, J. (1989). *An electromyographic study of preparatory set in singing as influenced by the Alexander technique.* Ph.D, Ohio State University.

Fishbein, M., Middlestadt, S., Ottati, V., Straus, S., and Ellis, A. (1988). Medical problems among ICSOM musicians:overview of a national survey. *Medical Problems of Performing Artists, 3*(1), 1-8.

French, S. N. (1978). Electromyographic biofeedback for tension control during gross motor skill acquisition. *Perceptual and motor skills, 47*(3), 883-889.

French, S. N. (1980). Electromyographic biofeedback for tension control during fine motor skill acquisition. [Comparative Study]. *Biofeedback and self-regulation, 5*(2), 221-228.

Garrison, T. (1978). *The effect of electromyographic biofeedback training on singers with tension problems in the laryngeal musculature.* D.A., Ball State University, Muncie. Retrieved from http://wwwlib.umi.com/dissertations/fullcit/7821104 (AAT 7821104)

Guettler, K. (1992). Electromyography and Muscle Activities in Double Bass Playing. *Music Perception: An Interdisciplinary Journal, 9*(3), 303-309.

Havas, K. (1961). *A New Approach to Violin Playing*. London: Bosworth and Co.

Heuser, F., and Mcnitt-Gray, J. L. (1994). Emg Changes in a Trumpet Player Overcoming Tone Commencement Difficulties - a Case-Study. *Medical Problems of Performing Artists, 9*(1), 18-24.

Heuser, F., and McNitt-Gray, J. L. (1998). Enhancing and validating pedagogical practice: The use of electromyography during trumpet instruction. *Medical Problems of Performing Artists, 13*(4), 155-159.

Hoppmann, R. (1998). Musculoskeletal problems in instrumental musicians. In R. Sataloff, A. Brandfonbrener and R. Lederman (Eds.), *Textbook of Performing Arts Medicine* (2 ed.). New York: Raven Press.

Iltis, P., and Givens, M. (2005). EMG Characterization of Embouchure Muscle Activity: Reliability and Application to Embouchure Dystonia. [Research]. *Medical Problems of Performing Artists, 20*(1), 25-33.

Irvine, J., and LeVine, W. R. (1981). The use of biofeedback to reduce left hand tension for string players. *American String Teacher, 31*, 10-12.

Kjelland, J. (2000). Application of Electromyography and Electromyographic Biofeedback in Music Performance Research: A Review of the Literature since 1985. *Medical Problems of Performing Artists, 15*(3), 115-118.

Koehler, W. (1995). *The Effect of Electromyographic Feedback on Achievement in Bowing Technique.* Paper presented at the Technological Directions in Music Education, San Antonio, TX.

Kolneder, W. (1993). *The Amadeus Book of the Violin - Construction, History, and Music.* Oregon: Amadeus Press.

Lederman, R. (1988). Occupational cramp in instrumental musicians. *Medical Problems of Performing Artists, 3*(2), 45-53.

Lehmann, A., and Davidson, J. (2002). Taking an acquired skills perspective. In R. Colwell and C. Richardson (Eds.), *The new handbook of research on music teaching and learning*. New York: Oxford University Press.

Levee, J. R., Cohen, M. J., and Rickles, W. H. (1976). Electromyographic biofeedback for relief of tension in the facial and throat muscles of a woodwind musician. [Case Reports]. *Biofeedback and self-regulation, 1*(1), 113-120.

LeVine, W. R. (1983). Behavioral and biofeedback therapy for a functionally impaired musician: a case report. *Biofeedback and self-regulation, 8*(1), 101-107.

LeVine, W. R., and Irvine, J. K. (1984). In vivo EMG biofeedback in violin and viola pedagogy. *Biofeedback and self-regulation, 9*(2), 161-168.

Levy, C., Lee, W., Brandfonbrener, A., Press, J., and Levy, A. (1992). Electromyographic analysis of muscular activity in the upper extremity generated by supporting a violin with and without a shoulder rest. *Medical Problems of Performing Artists, 7*(4), 103-110.

Metzner, P. (1998). *Crescendo of the Virtuoso: Spectacle, Skill and Self Promotion in Paris During the Age of Revolution*. London: University of California Press.

Middlestadt, S., and Fishbein, M. (1989). The prevelance of severe musculoskeletal problems among male and female symphony orchestra string players. *Medical Problems of Performing Artists, 4*(1), 41-48.

Montes, R., Bedmar, M., and Sol Martin, M. (1993). EMG biofeedback of the abductor pollicis brevis in piano performance. *Biofeedback and self-regulation, 18*(2), 67-77.

Moore, G., Hary, D., and Naill, R. (1988). Trills: Some initial observations. *Psychomusicology, 7*(2), 153-162.

Morasky, R. L., Reynolds, C., and Clarke, G. (1981). Using biofeedback to reduce left arm extensor EMG of string players during musical performance. *Biofeedback and self-regulation, 6*(4), 565-572.

Morasky, R. L., Reynolds, C., and Sowell, L. E. (1983). Generalization of lowered EMG levels during musical performance following biofeedback training. *Biofeedback and self-regulation, 8*(2), 207-216.

Niemann, B., Pratt, R., and Maughan, M. (1993). Biofeedback training, selected coping strategies, and music relaxation interventions to reduce debilitative musical performance anxiety. *International Journal of Arts Medicine,, 2*(2), 7-15.

Owen, E. (1985). Instrumental musicians and repetition strain injuries. *Journal of Occupational Health and Safety, Australia and New Zealand, 1*, 135-139.

Penn, I. W., Chuang, T. Y., Chan, R. C., and Hsu, T. C. (1999). EMG power spectrum analysis of first dorsal interosseous muscle in pianists. *Med Sci Sports Exerc, 31*(12), 1834-1838.

Reitman, A. (2001). The Effects of Music-assisted Coping Systematic Desensitization on Music Performance Anxiety. *Medical Problems of Performing Artists, 16*(3), 115-125.

Riley, k., and Chong, J. (2010). Garrick Olsson on Technique: mastering the paradoxical element of control. *Clavier Companion, 2*(5), 12-21.

Riley, K., Coons, E., and Marcarian, D. (2005). The Use of Multimodal Feedback in Retraining of Complex Technical Skills of Piano Performance. *Medical Problems of Performing Artists, 20*(2), 82-88.

Rosenkranz, K., Altenmuller, E., Siggelkowa, S., and Dengler, R. (2000). Alteration of sensorimotor integration in musician's cramp: impaired focusing of proprioception. *Clinical Neurophysiology, 111*(11), 2040-2045.

Ruiz, M. H., Jabusch, H. C., and Altenmüller, E. (2009). *New aspects in action planning and execution in musicians with dystonia.* Paper presented at the International Symposium on Performance Science 2009.

Samson, J. (2003). *Virtuosity and the Musical Work: The Transcendental Studies of Liszt*. New York: Cambridge University Press.

Shan, G., and Visentin, P. (2003). A quantitative three-dimensional analysis of arm kinematics in violin performance. *Medical Problems of Performing Artists, 18* (1), 3-10.

Shan, G., and Visentin, P. (2010). *Arts Biomechanics - An Infant Science: Its Challenges and Future*. New York: NOVA Publishers.

Shan, G., and Visentin, P. (2011). EMG applications in Studies of Arts Performance. In C. Steele (Ed.), *Electromyography/Book 2*. Rijeka: INTECH Publisher.

Steinmetz, A., Ridder, P., and Reichelt, A. (2006). Craniomandibular Dysfunction and Violin Playing: Prevalence and the Influence of Oral Splints on Head and Neck Muscles in Violinists. *Medical Problems of Performing Artists, 21*(4), 183-189.

Thiem, B., Greene, D., Perassas, S., and Thaut, M. (1994). Left arm muscle activation and movement patterns in cellists employing a playing technique using rhythmic cuing. *Medical Problems of Performing Artists, 9*(3), 89-96.

Visentin, P., and Shan, G. (2003). The kinetic characteristics of the bow arm during violin performance; an examination of internal loads as a function of tempo. *Medical Problems of Performing Artists, 18*(3), 91-98.

Visentin, P., Shan, G., and Schultz, A. (2004). Multi-dimensional Signal Analysis as a means of better understanding factors associated with repetitive use in violin performance. *Medical Problems of Performing Artists, 19*(3), 129-140.

Visentin, P., Shan, G., and Wasiak, E. B. (2008). Informing music teaching and learning using movement analysis technology. *International Journal of Music Education, 26*(1), 73-87.

Weber, M. J. (1995). *An investigation of selected muscle potential activity in violin/viola vibrato [microform]*. Ph.D, University of North Texas, Denton. Retrieved from http://worldcat.org/oclc/38363754

Wilson, F. (1986). *Tone deaf and all thumbs?* New York: Viking Press.

Yoshie, M., Kudo, K., and Ohtsuki, T. (2008). Effects of Psychological Stress on State Anxiety, Electromyographic Activity, and Arpeggio Performance in Pianists. *Medical Problems of Performing Artists, 23*(3), 120-132.

Zaza, C. (1998). Playing-related musculoskeletal disorders in musicians: A systematic review of incidence and prevalence. *Canadian Medical Association Journal, 158*(8), 1019-1025.

Zinn, M. (1998). *The use of sEMG feedback for transfer of relaxation into piano technique [unpublished]*. . Master, Illinois Institute of Technology, Chicago.

INDEX

E

F

G

H

N

O

P

Q

R

S

T